# METHODS IN MOLECULAR BIOLOGY



For further volumes:
http://www.springer.com/series/7651

For over 35 years, biological scientists have come to rely on the research protocols and methodologies in the critically acclaimed *Methods in Molecular Biology* series. The series was the first to introduce the step-by-step protocols approach that has become the standard in all biomedical protocol publishing. Each protocol is provided in readily-reproducible step-by-step fashion, opening with an introductory overview, a list of the materials and reagents needed to complete the experiment, and followed by a detailed procedure that is supported with a helpful notes section offering tips and tricks of the trade as well as troubleshooting advice. These hallmark features were introduced by series editor Dr. John Walker and constitute the key ingredient in each and every volume of the *Methods in Molecular Biology* series. Tested and trusted, comprehensive and reliable, all protocols from the series are indexed in PubMed.

# Magnaporthe oryzae

## Methods and Protocols

Edited by

**Stefan Jacob**

*Institute of Biotechnology and Drug Research gGmbH (IBWF), Mainz, Rheinland-Pfalz, Germany*

Humana Press

*Editor*
Stefan Jacob
Institute of Biotechnology and Drug
Research gGmbH (IBWF)
Mainz, Rheinland-Pfalz, Germany

ISSN 1064-3745 ISSN 1940-6029 (electronic)
Methods in Molecular Biology
ISBN 978-1-0716-1612-3 ISBN 978-1-0716-1613-0 (eBook)
https://doi.org/10.1007/978-1-0716-1613-0



This Humana imprint is published by the registered company Springer Science+Business Media, LLC, part of Springer Nature.
The registered company address is: 1 New York Plaza, New York, NY 10004, U.S.A.

# Preface

This volume highlights molecular methods to study the phytopathogenic rice blast fungus *Magnaporthe oryzae*. Over one-half of the world's population relies on rice as the main source of calories, representing the economic and scientific importance of this pathogen that has devastating effects on nutritional security. Apart from general informative sections about history, development, and evolution of the pathogen, molecular methods to increase the knowledge of the biology, genetic and metabolic diversity, as well as adaptability are presented in detail. *M. oryzae* has emerged as a major model system for understanding the molecular basis of fungal pathogenesis and the biology of plant/pathogen interaction.

***Mainz, Rheinland-Pfalz, Germany*** ***Stefan Jacob***

# Contents

## Contributors


TAKAYUKI ARAZOE • *Department of Applied Biological Science, Faculty of Science and Technology, Tokyo University of Science, Chiba, Japan*

KATHARINA BERSCHING • *Institute of Biotechnology and Drug Research gGmbH (IBWF), Mainz, Germany*

STEFAN BOHNERT • *Institute of Biotechnology and Drug Research gGmbH (IBWF), Mainz, Germany*

SONIA CHADHA • *Nuclear Agriculture and Biotechnology Division, Bhabha Atomic Research Centre, Mumbai, India; Homi Bhabha National Institute, Mumbai, India*

NEFTALY CRUZ-MIRELES • *The Sainsbury Laboratory, University of East Anglia, Norwich, UK*

MENG-DI DAI • *State Key Laboratory of Rice Biology, Institute of Biotechnology, Zhejiang University, Hangzhou, China*

NAWARAJ DULAL • *Department of Entomology and Plant Pathology, University of Arkansas Systems Division of Agriculture, Fayetteville, AR, USA*

MARTIN EGAN • *Department of Entomology and Plant Pathology, University of Arkansas Systems Division of Agriculture, Fayetteville, AR, USA*

IRIS EISERMANN • *The Sainsbury Laboratory, University of East Anglia, Norwich, UK*

MARISELA GARDUÑO-ROSALES • *The Sainsbury Laboratory, University of East Anglia, Norwich, UK; Department of Microbiology, Center for Scientific Research and Higher Education of Ensenada (CICESE), Ensenada, Mexico*

CIARÁN GRIFFIN • *School of Biological and Marine Sciences, University of Plymouth, Plymouth, UK*

CHRISTIANE GRÜNEWALD • *Johannes Gutenberg-University Mainz, Microbiology and Wine Research at the Institute for Molecular Physiology, Mainz, Germany*

STEFAN JACOB • *Institute of Biotechnology and Drug Research gGmbH (IBWF), Mainz, Germany; Johannes Gutenberg-University Mainz, Microbiology and Wine Research at the Institute for Molecular Physiology, Mainz, Germany*

GANG LI • *Department of Plant Pathology, University of Nebraska-Lincoln, Lincoln, NE, USA*

LIN LI • *State Key Laboratory of Rice Biology, Institute of Biotechnology, Zhejiang University, Hangzhou, China*

FU-CHENG LIN • *State Key Laboratory of Rice Biology, Institute of Biotechnology, Zhejiang University, Hangzhou, China*

GEORGE R. LITTLEJOHN • *School of Biological and Marine Sciences, University of Plymouth, Plymouth, UK*

XIAO-HONG LIU • *State Key Laboratory of Rice Biology, Institute of Biotechnology, Zhejiang University, Hangzhou, China*

YU LI • *State Key Laboratory of Rice Biology, Institute of Biotechnology, Zhejiang University, Hangzhou, China*

THOMAS MICHNA • *University Medical Center, Johannes-Gutenberg University Mainz, Mainz, Germany*

CAMILLA MOLINARI • *The Sainsbury Laboratory, University of East Anglia, Norwich, UK*

HENDRIK NEUMANN • *Johannes Gutenberg-University Mainz, Microbiology and Wine Research at the Institute for Molecular Physiology, Mainz, Germany*
ELY OLIVEIRA-GARCIA • *Department of Plant Pathology and Crop Physiology, Louisiana State University Agricultural Center, Baton Rouge, LA, USA*
RAQUEL O. ROCHA • *Department of Plant Pathology, University of Nebraska-Lincoln, Lincoln, NE, USA*
AUDRA ROGERS • *Department of Entomology and Plant Pathology, University of Arkansas Systems Division of Agriculture, Fayetteville, AR, USA*
LAUREN S. RYDER • *The Sainsbury Laboratory, University of East Anglia, Norwich, UK*
ELIZABETH SKELLAM • *Department of Chemistry & BioDiscovery Institute, University of North Texas, Denton, TX, USA*
LI-XIAO SUN • *State Key Laboratory of Rice Biology, Institute of Biotechnology, Zhejiang University, Hangzhou, China*
NICHOLAS J. TALBOT • *The Sainsbury Laboratory, University of East Anglia, Norwich, UK*
BOZENG TANG • *The Sainsbury Laboratory, University of East Anglia, Norwich, UK*
STEFAN TENZER • *University Medical Center, Johannes-Gutenberg University Mainz, Mainz, Germany*
BARBARA VALENT • *Department of Plant Pathology, Kansas State University, Manhattan, KS, USA*
RICHARD A. WILSON • *Department of Plant Pathology, University of Nebraska-Lincoln, Lincoln, NE, USA*
XIA YAN • *The Sainsbury Laboratory, University of East Anglia, Norwich, UK*
ALEXANDER YEMELIN • *Institute of Biotechnology and Drug Research gGmbH (IBWF), Kaiserslautern, Germany*
XUE-MING ZHU • *State Key Laboratory of Rice Biology, Institute of Biotechnology, Zhejiang University, Hangzhou, China*

# Chapter 1

# The Impact of Blast Disease: Past, Present, and Future

Barbara Valent

## Abstract

Rice blast disease is both the most explosive and potentially damaging disease of the world's rice (*Oryza sativa*) crop and a model system for research on the molecular mechanisms that fungi use to cause plant disease. The blast fungus, *Magnaporthe oryzae*, is highly evolved to sense when it is on a leaf surface; to develop a pressurized cell, the appressorium, to punch through the leaf cuticle; and then to hijack living rice cells to assist it in causing disease. Host specificity, determining which plants particular fungal strains can infect, is also an important topic for research. The blast fungus is a moving target, quickly overcoming rice resistance genes we deploy to control it, and recently emerging to cause devastating disease on an entirely new cereal crop, wheat. *M. oryzae* is highly adaptable, with multiple examples of genetic instability at certain gene loci and in certain genomic regions. Understanding the biology of the fungus in the field, and its potential for genetic and genome variability, is key to keep it from adapting to life in the research laboratory and losing relevance to the significant impact it has on global food security.

**Key words** *Magnaporthe oryzae*, *Pyricularia oryzae*, Rice blast, Wheat blast, Pathogenicity mechanisms, Host specificity, Sexual cycle, Parasexual cycle, Genome instability, Global food security

## 1 Introduction

### 1.1 Rice Blast in the Field

Rice blast, caused by the haploid, ascomyceteous fungus *Magnaporthe oryzae* (synonym of *Pyricularia oryzae*) [1], remains the most explosive and potentially damaging disease of the world's rice (*Oryza sativa*) crop [2]. An estimated $70 million USD annually is required for fungicide treatments to control blast disease in the Southern USA, where only 1 to 2% of global rice is grown [3]. Annual global loss due to rice blast is estimated at $66 billion, and enough grain is lost to feed 60 million people. *M. oryzae* is a seed-borne pathogen, providing an effective route for the global spread of the pathogen through movement of seed for planting and through international grain trade [4]. Strategies for durable, sustainable disease control remain elusive. Therefore, rice blast remains a threat to food security of approximately 3 billion people who depend on rice as their primary food, as well as a threat to export markets and economic livelihoods globally. Additionally,

Stefan Jacob (ed.), *Magnaporthe oryzae: Methods and Protocols*, Methods in Molecular Biology, vol. 2356,
https://doi.org/10.1007/978-1-0716-1613-0_1, 

**Fig. 1** Leaf and neck blast symptoms from the field. (**a**) Blast lesions at various stages of development on rice seedlings from natural inoculum in the blast nursery at the International Rice Research Institute, Philippines, in 1985. Susceptible lesions first become visible 4 to 5 days after conidia land on leaves, and they grow to their final size after ~7 days. Lesions begin as water-soaked spots that are first centered around the penetration site, and then expand over several days until stopped by brown bordering resistance reactions. (**b**) A sporulating eyespot lesion on a mature rice leaf in the Philippines. The light gray color reflects the color of the hyphae and conidiospores. (**c**) Neck blast symptoms from Japan in 2002. All grain development is blocked above the brown stem lesions. Note that seeds are developing below the point of neck infection. Also note the leaf lesion in the upper left corner. (Photos **a** and **c** from B. Valent; Photo **b** from J.M. Bonman)

rice blast introduces major unknowns into response to the impact of climate change on rice production [5–7].

The *M. oryzae Oryza* pathotype causes disease on all aboveground parts of the rice plant, with leaf blast, neck rot, and panicle blast responsible for significant losses in the field (Fig. 1). Leaf lesions developing from single appressorial penetration sites are typically eyespot- or spindle-shaped, and they are often bordered by dark brown margins surrounded by chlorotic halos (Fig. 1a, b). Sporulation capacity of an individual lesion depends on the relative area of the light tan center where sporulation occurs. Lesions range in size from ~1 mm in length (Type 2 lesions) through 5 to 10 mm in length (Type 5 lesions) depending on pathogen strain aggressiveness and the age and susceptibility of the host tissue [8]. Lesions appear gray during sporulation (Fig. 1b). Large susceptible lesions produce several thousand conidia every day for about 2 weeks [9]. Small uniformly dark brown spots without light centers (Type 1) fail to sporulate and represent macroscopically visible resistance reactions that usually appear within 2 to 3 days [8]. Infection of the highly susceptible panicle neck (Fig. 1c) blocks all seed development above the infection point (Fig. 1c) and represents an especially devastating form of blast disease.

Current understanding of the early asymptomatic infection biology is detailed in Chapter 2, including development and function of appressoria that build up high levels of turgor pressure to mechanically puncture through the plant surface. Subsequent tissue invasion involves specialized intracellular invasive hyphae that enter and feed on living host cells, one after another, with no visible macroscopic symptoms for 4 to 5 days after the spore lands on the host surface [10–14]. Understanding this biotrophic invasion strategy defined a new paradigm in hemibiotrophy in which the fungus invades and feeds on live plant cells, but each invaded cell dies by the time the fungus moves into the living neighbor cells [10, 11]. The earliest indication of macroscopic symptoms are faint discolored water-soaked spots that grow into sporulating lesions. This visible necrotrophic phase of lesion development appears to involve the expression of plant cell wall-degrading enzymes. It is unknown what signals trigger the necrotrophic transition to macroscopic symptoms and sporulation.

## 1.2 Additional Blast Diseases with Major Impact

As important as rice blast is, the entire *M. oryzae* disease complex poses an even greater threat to global food security. Blast diseases of additional cereal crops are caused by different crop-adapted lineages, called pathotypes, of *M. oryzae* [15–17]. Blast diseases limit production of millets, subsistence crops that have been grown for thousands of years by low-resource farmers in Africa and Asia. Finger millet (*Eleusine coracana* (L.) Gaertn. subsp. *coracana*), which is still grown by millions of smallholder farmers in Eastern Africa and Southern India [18], is relatively disease-free except for blast disease caused by the *M. oryzae Eleusine* pathotype [17, 19]. Foxtail (Italian) millet (*Setaria italica*) was perhaps the first domesticated grain crop, and it is still grown by smallholder farmers in East Asia [20]. Blast caused by the *Setaria* pathotype is the major disease on foxtail millet. In addition to these ancient cereal diseases, two serious new blast diseases emerged within the last 35 years. Blast disease on wheat (*Triticum aestivum*) is caused by the *M. oryzae Triticum* pathotype, and blast on *Lolium* and *Festuca* forage and turf grass crops is caused by the *M. oryzae Lolium* pathotype. Wheat blast was first identified in 1985 in Paraná State of Brazil and it rapidly spread, causing recurring epidemics in Brazil, Bolivia, and Paraguay [4, 21]. Wheat blast moved into Bangladesh in 2016, with devastating consequences [22, 23]. *Lolium* blast emerged in the USA in the early to mid-1990s as a serious disease on turf and forage grasses [24, 25], including perennial ryegrass (*Lolium perenne*), annual ryegrass (*Lolium multiflorum*), and tall fescue (*Festuca arundinacea*). The disease has now spread to Japan, China, and France, presumably via infected seed [26–28]. The *Triticum*-adapted and *Lolium*-adapted lineages are closely related and highly diverse compared to the much older lineages adapted to rice and the millets [15]. Thus, *M. oryzae* represents moving target for disease control, but research on these other serious blast diseases lags behind.

## 2 Conidia, Ascospores, Microconidia, and Other Biology

### 2.1 Conidia and the Disease Cycle

The blast fungus executes diverse developmental processes, including the best-studied hyphal morphogenesis associated with its asexual disease cycle (*see* Chapter 2). The hyaline to pale gray-colored asexual conidia each contain three cells, with a single identical nucleus per cell (Fig. 2a). Therefore, the fungus is routinely purified by isolation of a single conidium [29]. Conidiogenesis in *M. oryzae* is holoblastic such that expansion and swelling of the conidiophore apex gives rise to a conidium, followed by a septum being formed to delimit the forming conidium [30]. The apex then grows to the side to produce the next conidium, resulting in four or more conidia borne sympodially on a conidiophore.

### 2.2 The Elusive Sexual Cycle

The *M. oryzae* sexual cycle has not been observed in nature, but some strains undergo sexual crosses in the laboratory [31–34]. The fungus is a Pyrenomycete, producing ascospores in unordered asci within perithecia (Fig. 2b, c). Fully fertile strains are self-sterile hermaphrodites (functioning both as females and males), with compatibility for mating governed by two alternate alleles of the mating type locus *MAT1*. Strains of opposite mating type mate to form pigmented perithecia (500–1200 μm in length) with spherical bases (80–260 μm in diameter) and long cylindrical necks. Unordered asci contain 8 hyaline, crescent-shaped ascospores (16–25 × 4–8 μm), each with 4 cells containing a single mitotically derived nucleus. Perithecia produce asci within 2 to 3 weeks when strains of opposite mating type are incubated on oatmeal agar in the light at ~20 °C [29]. Mature asci are extruded through the perithecial necks and ascospores are released into a viscous liquid which can solidify on the perithecial neck [8]. Ascospores produce appressoria for plant penetration (Fig. 2b), and they disperse through vegetative growth and conidiation.

The degree of sexual fertility exhibited by fungal isolates in different host-specific populations within *M. oryzae* ranges from highly fertile hermaphrodites to female-sterile strains that mate only with hermaphrodites of opposite mating type, to totally infertile strains. Host range is generally predictive of fertility. For rice pathogens, low levels of sexual fertility and geographic isolation of strains with a single mating type suggested that most field populations infecting rice were predominantly asexual [35, 36]. Rare populations in Southeast Asia, the presumed center of origin of the pathogen, show signatures of sexual reproduction [37, 38]. In contrast, most *Eleusine* isolates [17, 35] and *Triticum* isolates from the 1980s and early 1990s are sexually fertile hermaphrodites that cross to produce abundant viable ascospores [39, 40]. Guy11, a rare hermaphroditic rice isolate from French Guyana, is a commonly used wild-type strain for rice pathogen research

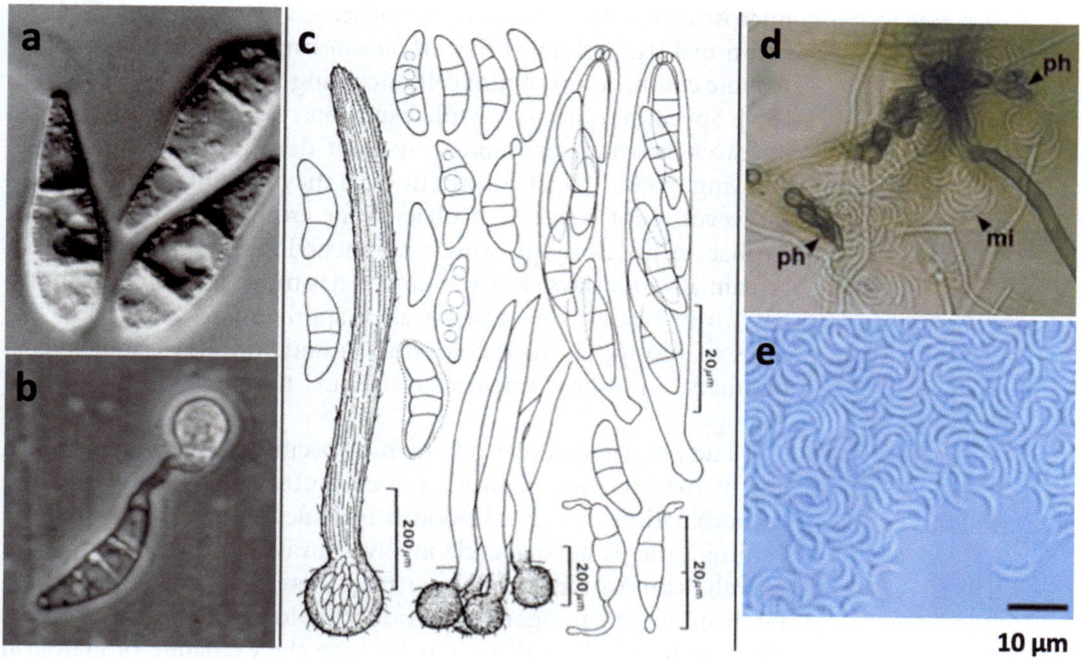


**Fig. 2** Three types of spores produced by the blast fungus. (**a**) Asexual pyriform conidia (25–30 × 9–12 μm) produce germ tubes that differentiate appressoria on inductive surfaces. (Reproduced with permission from Hamer et al. [68]). (**b**) Crescent-shaped four-celled ascospores also form appressoria (B. Valent). (**c**) Depiction of perithecia and ascospores, reproduced with permission from Yaegashi and Udagawa [33]. (**d**) Asexual microconidia (mi) are produced from phialides (ph). (**e**) Close-up view of microconidia. (Images **d** and **e** reproduced with permission from Chuma et al. [50])

[41, 42]. Guy11 was crossed with *Eleusine* pathogens for improved fertility, resulting in laboratory strain 70-15 that was used for the reference genome [29, 43, 44]. Both tetrad analysis and random spore analysis have identified multiple genes involved in *M. oryzae* host specificity [45–49].

Many sexually fertile isolates also produce small, crescent-shaped microconidia in phialides (Fig. 2d, e). These spores have been hypothesized to function as spermatia, but there is still no evidence for this function [50, 51]. Recently, microconidia have been observed to germinate and infect plants through wounds, but their role in nature is unknown [52].

### 2.3 Other Biology—Root Infection and the Parasexual Cycle

Production of microconidia from phialides by fertile *M. oryzae* strains is consistent with the fungus's close taxonomic relatedness to the phialide-producing root pathogens of grass hosts, *Magnaporthe poae* and *Gaeumannomyces graminis* [53]. In laboratory studies, *M. oryzae* is able to infect roots using common tactics of these root-invading relatives. These include forming thick melanized runner hyphae on roots, as well as forming hyphopodia, infection pads that produce infection pegs to penetrate the root surface. Putative swollen intracellular resting structures and microsclerotia-like structures are observed in root tissue [54]. An additional study showed that *M. oryzae* can execute extended biotrophic colonization of root cells including the root vascular system [55]. Specifically, invasive hyphae in roots resemble invasive hyphae in leaves in being surrounded by plant-derived membrane and in showing major hyphal constrictions as they pass into the new cells. However, root invasive hyphae differ from biotrophic invasive hyphae in leaves by growing unbranched through rice cells that remain alive for days after the fungus has moved on. In the laboratory, root disease can develop after approximately 2 weeks. The roles of biotrophic root colonization and root-associated hyphal plasticity, including microsclerotia, are unknown in blast field biology.

The parasexual cycle is known to occur in *M. oryzae* [56–59], and it may provide a means for chromosome reassortment and recombination in asexual populations such as the rice pathogen. In fungi, the parasexual cycle involves fusion of hyphae from vegetatively compatible strains, forming heterokaryotic cells which rarely undergo karyogamy to produce diploid cells. Mitotic recombination in the diploid nucleus leads to the exchange of genes on chromosomes. *M. oryzae* diploid strains are unstable, and they rapidly revert to haploids through random loss of one of each pair of homologous chromosomes [56, 60]. Therefore, gene assortment occurs through mitotic recombination and random reassortment of the chromosomes from the two parents. The impact of parasexual recombination on evolution of *M. oryzae* populations in nature requires further research.

## 3 From the Field into the Laboratory

### 3.1 M. oryzae *Is Unstable During Subculture on Artificial Media*

Understanding the fungus in its natural habitat requires maintaining the fungus in its natural state from the field. *M. oryzae* strains are unstable during continuous subculture with regard to morphology, sexual fertility, conidiation, and pathogenicity [29, 61]. One characteristic associated with many fresh field isolates is "mosaic" morphology on oatmeal or rice polish agar generally used for genetic crosses (Fig. 3). Mosaic cultures derived from a single conidium are at first smooth and heavily conidiated. Approximately 10 days after conidiation, papillae develop on top of the previously established mycelium. The papillae can take the form of heavily conidiated clumps or of non-conidiating mycelial outgrowths. Drops of an extremely viscous ooze also appear on these cultures. The mosaic trait appears to be easily lost, often within a few rounds of subculture on agar plates. When these mosaic strains are subcultured by mass hyphal transfers on oatmeal agar, a progression of morphological changes occurs. The cultures cease to papillate, lose their normal gray pigment, and conidiate poorly. They show a

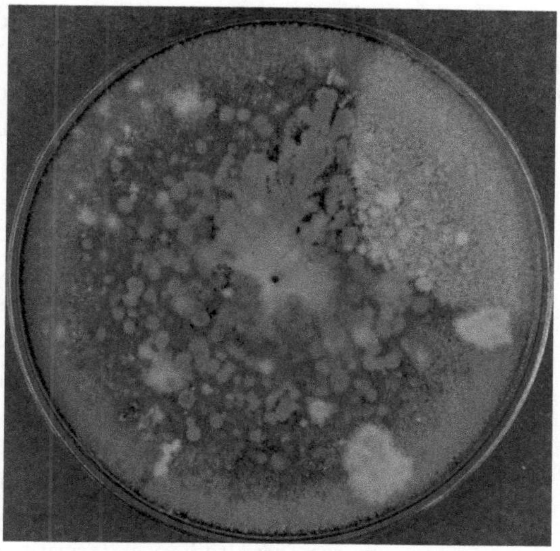

**Fig. 3** Typical "mosaic" morphology of fresh field isolates on standard oatmeal agar. Papillae with different hyphal or sporulation morphologies develop on the surface of established mycelial cultures. Transfer of papillae leads to cultures with changed morphology. This plate also shows a spontaneous Buf$^{-}$ (orange) sector, with a mutation in the *BUF1* gene involved in melanin biosynthesis. (Photo from B. Valent)

phenomenon known as "melting out," which appears to involve autolysis of mycelium that is more than 10–14 days old. This degraded condition never reverts to mosaic morphology. Loss of fertility, specifically the loss of the ability to serve as a female in a cross, occurs as rapidly upon subculture as loss of the mosaic morphology. Recently, experimental evolution experiments have demonstrated genetically stable loss of sexual fertility after 10 to 19 rounds of weekly transfer of conidia to new agar plates [62], showing that the instability is not just a function of mycelial subculture. The ability of a strain to serve as a male in a cross appears to be relatively more stable. Although pathogenicity is less sensitive to subculturing than female fertility, loss of pathogenicity does occur, and the loss of the ability to produce abundant conidia seriously restricts pathogenicity testing. Thus, it is critical to store all strains in a non-metabolizing state and to recover strains from storage directly before experiments. Routine storage involves fungus that is dried and frozen in filter paper discs or pieces, with backup storage in the form of dried and frozen infected plant material [29, 63]. In the Valent lab, diverse fungal strains that were dried and frozen in filter papers or in infected host tissue in 1986 through 1987 are still viable. Using desiccants to prevent the fungus from regaining moisture during frozen storage is key to long-term preservation of *M. oryzae* strains.

### 3.2 Examples of Genome Instability in Field Isolates

Genetic instability of rice pathogens is clearly documented by the rapid spontaneous mutation of certain genes involved in pathogenicity and host specificity. For example, some rice pathogens undergo frequent spontaneous deletion of the *BUF1* melanin-biosynthesis gene encoding 1,3,8-trihydroxynaphthalene reductase. This single-copy gene is essential for appressorial pressure buildup, host penetration, and pathogenicity [30, 64]. Mutants with spontaneous deletion of the *BUF1* gene are observed as sectors or as orange papillae against the gray wild-type fungus on agar plates (Fig. 3). Some rice isolates produce up to 0.25% *buf*$^-$ conidia during vegetative growth [64] or cross with other wild-type rice isolates to produce from 4 to 20% *buf*$^-$ ascospore progeny [8, 64]. Molecular identification of *BUF1* gene [65] showed that the gene was being deleted at these high frequencies. Detailed genetic analysis of instability of mutable *BUF1* alleles in sexual crosses resulted in a description of meiosis-associated deletion in heteroallelic repeats (MDHR), a phenomenon in which instability of the *BUF1* chromosomal region is strongly influenced by its homologous pairing partner [66]. This study demonstrates that investigating specific examples of blast genome instability can raise questions about long-assumed dogma based on work with other fungi, in this case suggesting that karyogamy might precede DNA replication in the blast fungus or that parts of the genome remain unreplicated until after karyogamy and subsequent chromosome pairing.

Examples of genetic instability have been associated with signal transduction components. Null mutations in the *MAC1* gene encoding adenylate cyclase have been produced using gene knockout techniques, and these mutants exhibit similar unstable mosaic growth with frequent sectoring on oatmeal agar plates as shown in Fig. 3 [67]. Loss of *MAC1*, involved in cAMP signaling, shows pleiotropic effects on growth, conidiation, sexual development, and appressorium formation. Characterization of frequent spontaneous wild-type-appearing sectors on agar plates identified *mac1*$^-$ revertants containing second-site suppressor mutations, including one (sum1-99) that alters a conserved leucine in a cAMP binding domain of the regulatory subunit of the cAMP-dependent protein kinase A. In a second example, early rice blast studies were characterized by frequent spontaneous mutants with abnormally shaped conidia and appressoria when screening for other phenotypes [68]. Genetic analyses of these mutants identified a single gene named *SMO1* for spore morphology. *SMO1* was a challenge to clone, and it has only recently been identified as encoding a Ras GTPase-activating protein involved in spore morphology and septin-mediated plant infection [69]. In contrast to frequent deletion of the *BUF1* gene, spontaneous mutational events leading to inactivation of *SMO1* were single-nucleotide polymorphisms

(SNPs) or small insertions or deletions [69]. The mechanism for frequent occurrence of single-nucleotide polymorphisms in the cases of *mac1*$^-$ suppressors and *SMO1* is unknown.

Another trait in which host-specific populations differ is telomere instability. Instability of telomere sequences in rice isolates has been documented, and indeed, about half of the known *AVR* effector genes are located near telomeres [45, 70]. Interestingly, *Lolium* isolates show unprecedented levels of telomere instability relative to rice isolates [71]. Mechanisms and biological consequences for telomere instability traits in the host-specific field populations remain interesting subjects for analysis. However, these key examples of genome instability in *M. oryzae* underscore the need to minimize subculturing during laboratory manipulation.

## 4 The Genetics and Genomics of Host Specificity

### 4.1 Rice Cultivar Specificity

For rice blast, differences in rice cultivar specificity were first distinguished by Sasaki in Japan in 1922, and he also reported the first genetic analysis of resistance in rice [72]. Since that time hundreds of races have been defined by their virulence spectrums on differential rice cultivars containing different resistance (*R*) genes [2]. Major effort has been directed toward identifying dominant resistance *R* genes in rice or wild relatives and toward developing blast-resistant cultivars. This has resulted in the identification of ~100 *R* genes that prevent disease by some pathogen races, as well as ~350 quantitative trait loci [73, 74]. Despite this wealth of *R* genes, genetic control has not been effective, due to the variability of the fungus and its ability to "defeat" introduced *R* genes within 1 to 2 years of their deployment in the field.

Rice blast is a classical "gene-for-gene" system in which resistance occurs when *R* gene products directly or indirectly recognize pathogen *AVR* gene products [75]. *AVR* genes generally encode small secreted effector proteins that are specifically expressed during biotrophic invasion. For the rice blast fungus, 12 *AVR* genes have been molecularly characterized, including 11 corresponding to characterized *R* genes [74]. Insights into the mechanisms by which blast AVR effectors trigger rice *R* gene-mediated resistance are discussed in Chapter 2 and further detailed in reviews [74, 76, 77].

### 4.2 Host Species Specificity

Multiple examples of single genes that confer host specificity at the plant genus (species) level have been described since the original report by Yaegashi in 1978 [78]. These include *PWL1* from finger millet pathogens and *PWL2* from rice pathogens [46, 79], which prevent individual strains from infecting weeping lovegrass (*Eragrostis curvula*). Tosa and colleagues have used genetic crosses to confirm a gene-for-gene relationship underlying the incompatibility of strains of various *M. oryzae* pathotypes with wheat [47, 80]. They identified five *AVR* effector genes that independently block infection of wheat, including *PWT1* and *PWT2* from both rice and *Setaria* pathogens [48, 81], *PWT3* and *PWT4* genes from an oat pathogen [80], and *PWT5* from a rice pathogen [48]. Indeed, loss of the *PWT3* gene/*Rwt3* gene interaction apparently contributed to the emergence of wheat blast in Brazil [82]. Specifically, wheat varieties planted before the 1980s contained the *R* gene *Rwt3*, which blocked *Lolium* and oat isolates with the corresponding *PWT3* gene from infecting wheat. In the 1980s, new *rwt3*$^-$ wheat varieties lacking this *R* gene were planted in a new blast-conducive region in Brazil during a prolonged period of wet, warm El Niño-fueled weather [21]. Subsequent mutation to loss of *PWT3* function allowed the emerging *Triticum* population to cause disease on all wheat varieties.

Not all differences in host species specificity are due to *AVR*-like genes. One study showed polygenic segregation of genes from a finger millet pathogen controlling lesion size in rice, as well as *AVR* genes corresponding to different rice cultivars [8]. This relatively unexplored aspect of host species specificity may contribute to adaptation to enhanced aggressiveness toward a particular host once all *AVR* gene obstacles to infection are removed.

### 4.3 AVR Gene Dynamics Impacts Disease Control

Newly assembled reference genomes for the rice isolate Guy11 [83], *Triticum* isolate B71 [84], and *Eleusine* isolate MZ5-1-6 [85] show a high degree of macrosynteny among the seven core chromosomes between these different host-adapted field isolates and the *M. oryzae* genome reference laboratory strain 70-15 [44]. These three sexually fertile field isolates share a common chromosome structure that identifies a major translocation involving chromosomes 1 and 6 that appears specific for the 70-15 (MG8) reference genome. *M. oryzae* genomes typically contain 10–13% repetitive sequences including active transposable elements (TE), which are implicated in chromosome rearrangements and gene gain and loss events [86, 87]. Individual rice isolates from asexual populations that have lost ability to complete the sexual cycle accumulate more chromosome rearrangements, presumably due to them being released from selection for homologous chromosome pairing during meiosis [39, 88].

Genome sequencing of the *Triticum* isolate B71 [84] provided sequence assemblies for small dispensable supernumerary chromosomes, called mini-chromosomes, that are found in some, but not all, *M. oryzae* field isolates [39, 88]. Mini-chromosomes appear highly variable, with the B71 mini-chromosome containing 52.8% repetitive sequences compared to 9.7% in the core chromosomes. Three mini-chromosomes from two different *Triticum* isolates contain shared and different *AVR* effector homologues, putative effector genes, and other sequences found on core chromosome ends in other strains, suggesting a cross-talk between these two chromosomal compartments [84]. These results suggest that dispensable mini-chromosomes and core chromosome ends are coupled as a mobile, fast-evolving effector compartment compared to slower-evolving regions of the core chromosomes in a classical two-speed genome [84, 89].

The rice blast fungus quickly overcomes *R* genes deployed for disease control. This often occurs by deletion of the corresponding *AVR* genes. However, *AVR* gene inactivation through transposon insertion and through nucleotide substitutions that alter recognition specificity also occurs [45, 46, 90]. Genome sequencing of diverse field isolates shows that effector genes, and in particular *AVR* effector genes, show tremendous loss-gain polymorphism [86, 87]. For example, frequent spontaneous deletions of the rice pathogen *AVR-Pita1*, in addition to some nucleotide substitutions and TE inactivation events, lead to loss of avirulence activity in rice varieties with the corresponding *Pita* gene in laboratory studies [45, 91] and in the field [92–94]. In diverse rice isolates, *AVR-Pita1* was absent or present in one or more copies, often near telomeres on chromosomes 1, 3, 5, or 7, and supernumerary mini-chromosomes, as well as at three different locations on chromosome 4. In *Setaria* isolates, *AVR-Pita1* homologues reside near a telomere on chromosome 6 or on a supernumerary chromosome. In contrast to the chromosomal mobility of the functional *AVR-Pita1* gene, *avr-pita3* lacks AVR function and it is stably located on chromosome 7 in these same strains. Chuma et al. [95] suggest that this extraordinary variability in *AVR-Pita1* location in rice isolates is due to loss and recovery of the *AVR* gene in response to periodic deployment of *Pita* in rice. That is, the deleted *AVR* gene is retained in the pathogen population and can be recovered by individual strains that lack it, although at different chromosomal locations. The parasexual cycle would provide a means for transfer of *AVR* genes between strains in asexual populations. Other *AVR* genes show signs of chromosome mobility, but some appear more stably located in the genome. Understanding the impact of genome location on *AVR* effector gene dynamics and *R* gene durability should be a high priority for future research.

## 5 Future Prospects

In the last three decades, rice blast has become a model system for understanding the molecular mechanisms of plant infection by a fungal pathogen [30, 96–104]. As described in Chapter 2, we have learned much about how the highly pressurized appressorium develops and functions to penetrate the plant surface. However, many details remain to be described. We are now only beginning to understand how specialized biotrophic invasive hyphae grow from cell to cell while flooding the rice tissue with hundreds of small effector proteins and other molecules to block defenses and hijack plant processes for its own purpose. Major questions remain on how the fungus so precisely controls the biotrophic invasion-specific gene expression of these effector genes and how the effectors function to promote disease. Finally, little is known about the later stage of lesion development when the tissue visibly dies and sporulation occurs. Major questions remain about the elaborate hyphal morphogenic transitions, environmental sensing, signaling, and highly coordinated gene expression that characterize all phases of the disease cycle. Understanding pathogenicity mechanisms provides the opportunity to intervene by designing host plants or disease management practices that present the fungus with an obstacle it is not able to overcome. That is, we need to identify the fungus's vulnerabilities during the infection cycle and find a way to stop it.

Breakdown of blast resistance is still a major cause of yield instability in major rice-growing areas [6]. It is critical to understand both the molecular mechanisms of *AVR* effector recognition and resulting resistance [74, 76, 105, 106], as well as the genetic mechanisms that allow the fungus to rapidly defeat the deployed resistance [2, 74, 90]. AVR effectors that contribute to pathogen fitness have been predicted to correspond to durable *R* genes for some diseases [107]. However, laboratory studies with *M. oryzae* have failed to identify individual *AVR* effectors with a measurable effect on pathogenicity. Additionally, field population analyses confirm a high degree of gain/loss polymorphism for *AVR* effector genes, indicating they are dispensable to the pathogen. This suggests a high degree of functional redundancy among *M. oryzae* effectors such that losing one has little impact on the fungus's ability to cause disease. Understanding *AVR* effector functions in the disease process would enable targeting sets of effectors with overlapping functions, for example, with *R* gene pyramids in rice, in order to lessen the fungus's ability to overcome resistance without a fitness penalty. It is critical to understand how genome location of *AVR* effector genes impacts the pathogen's ability to overcome resistance. There is currently no evidence that *R* genes corresponding to effector genes in the more stable core regions of the genome are more durable than *R* genes corresponding to genes that reside in the variable genome compartment comprising core chromosome ends and supernumerary chromosomes [84, 93, 95]. But this needs further investigation.

Phylogenetic analysis indicates that the rice-infecting population originated from a host shift from a *Setaria* population around the time of domestication of both rice and foxtail millet ~7000 years ago in the middle of Yangtze valley in China [108]. The first reports of rice cultivar specificity for the fungus and of genetics of rice resistance by Sasaki in 1922 initiated a century of breeding and research to control this disease [72]. In contrast to rice, few effective *R* genes have been identified for the other blast diseases, and little is known about disease mechanisms [4, 109]. The foxtail and finger millet diseases, while important for centuries in certain communities, are generally grown by smallholder farmers with less breeding and research directed at disease control. *Triticum* and *Lolium* isolates responsible for the wheat and ryegrass diseases, dating from 30 to 40 years ago, still appear to be in the process of gaining aggressiveness and adapting to their respective hosts. The different evolutionary stages of the various blast pathotypes provide an opportunity to study the stages and mechanisms of host adaptation. Detailed understanding of rice blast will guide research on understanding other host-adapted forms of *M. oryzae* as well as other fungal pathogens that threaten global agriculture and food security.

## References


1. Zhang N, Luo J, Rossman AY, Aoki T, Chuma I, Crous PW, Dean R, de Vries RP, Donofrio N, Hyde KD, Lebrun M-H, Talbot NJ, Tharreau D, Tosa Y, Valent B, Wang Z, Xu J-R (2016) Generic names in Magnaporthales. IMA Fungus 7:155–159
2. Wang G-L, Valent B (2009) Advances in genetics, genomics and control of rice blast disease. Springer Science and Business Media, New York, New York
3. Nalley L, Tsiboe F, Durand-Morat A, Shew A, Thoma G (2016) Economic and environmental impact of rice blast pathogen (*Magnaporthe oryzae*) alleviation in the United States. PLoS One 11:e0167295
4. Cruz CD, Valent B (2017) Wheat blast disease: danger on the move. Trop Plant Pathol 42:210–222
5. Asibi AE, Chai Q, Coulter JA (2019) Rice blast: a disease with implications for global food security. Agronomy 9:451
6. Khush GS, Jena KK (2009) Current status and future prospects for research on blast resistance in Rice (*Oryza sativa* L.). In: Wang GL, Valent B (eds) Advances in genetics, genomics and control of rice blast disease. Springer, Netherlands, pp 1–10
7. Skamnioti P, Gurr SJ (2009) Against the grain: safeguarding rice from rice blast disease. Trends Biotechnol 27:141–150
8. Valent B, Farrall L, Chumley FG (1991) *Magnaporthe grisea* genes for pathogenicity and virulence identified through a series of backcrosses. Genetics 127:87–101
9. Kato H (1974) Epidemiology of rice blast disease. Rev Plant Prot Res 7:1–20
10. Berruyer R, Poussier S, Kankanala P, Mosquera G, Valent B (2006) Quantitative and qualitative influence of inoculation methods on *in planta* growth of rice blast fungus. Phytopathology 96:346–355
11. Kankanala P, Czymmek K, Valent B (2007) Roles for rice membrane dynamics and plasmodesmata during biotrophic invasion by the blast fungus. Plant Cell 19:706–724
12. Khang CH, Berruyer R, Giraldo MC, Kankanala P, Park S-Y, Czymmek K, Kang S, Valent B (2010) Translocation of *Magnaporthe oryzae* effectors into rice cells and their subsequent cell-to-cell movement. Plant Cell 22:1388–1403
13. Koga H (2001) Cytological aspects of infection by the rice blast fungus *Pyricularia oryzae*. In: Sreenivasaprasad S, Johnson R (eds) Major Fungal Disease of Rice: Recent Advances. Kluwer Academic Publishes, Dordrecht, pp 87–110
14. Koga H, Dohi K, Nakayachi O, Mori M (2004) A novel inoculation method of *Magnaporthe grisea* for cytological observation of the infection process using intact leaf sheaths of rice plants. Physiol Mol Plant Pathol 64:67–72
15. Gladieux P, Condon B, Ravel S, Soanes D, Maciel JLN, Nhani A, Chen L, Terauchi R, Lebrun M-H, Tharreau D, Mitchell T, Pedley KF, Valent B, Talbot NJ, Farman M, Fournier E (2018) Gene flow between divergent cereal-

and grass-specific lineages of the rice blast fungus *Magnaporthe oryzae*. mBio 9: e01219–e01217

16. Valent B, Farman M, Tosa Y, Begerow D, Fournier E, Gladieux P, Islam MT, Kamoun S, Kemler M, Kohn LM, Lebrun M-H, Stajich JE, Talbot NJ, Terauchi R, Tharreau D, Zhang N (2019) Pyricularia graminis-tritici is not the correct species name for the wheat blast fungus: response to Ceresini et al. (MPP 20:2). Mol Plant Pathol 20:173–179
17. Kato H, Yamaguchi T, Nishihara N (1977) Seed transmission, pathogenicity and control of ragi blast fungus and susceptibility of ragi to *Pyricularia* spp. from grasses, cereals and mioga. Ann Phytopathol Soc Japan 43:392–401
18. Lenné JM, Takan JP, Mgonja MA, Manyasa EO, Kaloki P, Wanyera N, Okwadi J, Muthumeenakshi S, Brown AE, Tamale M, Sreenivasaprasad S (2007) Finger millet blast disease management - a key entry point for fighting malnutrition and poverty in East Africa. Outlook Agric 36:101–108
19. Takan JP, Akello B, Esele P, Manyasa EO, Obilana AB, Audi PO, Kibuka J, Odendo M, Oduori CA, Ajanga S, Bandyopadhyay R, Muthumeenakshi S, Coll R, Brown AE, Talbot NJ, Sreenivasaprasad S (2004) Finger millet blast pathogen diversity and management in East Africa: a summary of project activities and outputs. International Sorghum and Millets Newsletter 45:66–69
20. Sharma R, Girish AG, Upadhyaya HD, Humayun P, Babu TK, Rao VP, Thakur RP (2013) Identification of blast resistance in a core collection of foxtail millet germplasm. Plant Dis 98:519–524
21. Igarashi S (1990) Update on wheat blast (Pyricularia oryzae) in Brazil, abstr wheat for the nontraditional warm areas: a proceedings of the international conference, Foz Do Iguaçu, Brazil, July 29- august 3, 1990
22. Islam MT, Croll D, Gladieux P, Soanes DM, Persoons A, Bhattacharjee P, Hossain MS, Gupta DR, Rahman MM, Mahboob MG, Cook N, Salam MU, Surovy MZ, Sancho VB, Maciel JLN, Nhani A, Castroagudin VL, Reges JTD, Ceresini PC, Ravel S, Kellner R, Fournier E, Tharreau D, Lebrun MH, McDonald BA, Stitt T, Swan D, Talbot NJ, Saunders DGO, Win J, Kamoun S (2016) Emergence of wheat blast in Bangladesh was caused by a south American lineage of *Magnaporthe oryzae*. BMC Biol 14:84
23. Malaker PK, Barma NCD, Tiwari TP, Collis WJ, Duveiller E, Singh PK, Joshi AK, Singh RP, Braun HJ, Peterson GL, Pedley KF, Farman ML, Valent B (2016) First report of wheat blast caused by *Magnaporthe oryzae* pathotype triticum in Bangladesh. Plant Dis 100:2330–2330
24. Farman ML (2002) *Pyricularia grisea* isolates causing gray leaf spot on perennial ryegrass (Lolium perenne) in the United States: relationship to *P. grisea* isolates from other host plants. Phytopathology 92:245–254
25. Viji G, Wu B, Kang S, Uddin W, Huff DR (2001) *Pyricularia grisea* causing gray leaf spot of perennial ryegrass turf: population structure and host specificity. Plant Dis 85:817–826
26. Milazzo J, Pordel A, Ravel S, Tharreau D (2018) First scientific report of *Pyricularia oryzae* causing gray leaf spot disease on perennial ryegrass (*Lolium perenne*) in France. Plant Dis 105:1024
27. Tosa Y, Hirata K, Tamba H, Nakagawa S, Chuma I, Isobe C, Osue J, Urashima AS, Don LD, Kusaba M, Nakayashiki H, Tanaka A, Tani T, Mori N, Mayama S (2004) Genetic constitution and pathogenicity of *Lolium* isolates of *Magnaporthe oryzae* in comparison with host species-specific pathotypes of the blast fungus. Phytopathology 94:454–462
28. Xue LH, Liu Y, Li CJ, Wu WX (2017) First report of gray leaf spot (*Pyricularia oryzae*) on Italian ryegrass (*Lolium multiflorum*) in China. Plant Dis 101:1049–1049
29. Valent B, Crawford MS, Weaver CG, Chumley FG (1986) Genetic studies of fertility and pathogenicity in *Magnaporthe grisea*. Iowa State J Res 60:569–594
30. Howard RJ, Valent B (1996) Breaking and entering: host penetration by the fungal rice blast pathogen *Magnaporthe grisea*. Annu Rev Microbiol 50:491–512
31. Kato H, Yamaguchi T, Nishihara N (1976) The perfect state of *Pyricularia oryzae* Cav. In culture. Japanese J Phytopathol 42:507–510
32. Yaegashi H, Hebert TT (1976) Perithecial development and nuclear behavior in *Pyricularia*. Phytopathology 66:122–126
33. Yaegashi H, Udagawa S (1978) Taxonomical identity of perfect state of *Pyricularia grisea* and its allies. Can J Bot 56:180–183
34. Ueyama A, Tsuda M (1976) Formation of the perfect state in culture of *Pyricularia* sp. from some graminaceous plants (preliminary report). Trans Mycol Soc Jpn 16:420–422
35. Takan JP, Chipili J, Muthumeenakshi S, Talbot NJ, Manyasa EO, Bandyopadhyay R,

Sere Y, Nutsugah SK, Talhinhas P, Hossain M, Brown AE, Sreenivasaprasad S (2012) *Magnaporthe oryzae* populations adapted to finger millet and rice exhibit distinctive patterns of genetic diversity, sexuality and host interaction. Mol Biotechnol 50:145–158

36. Zeigler RS, Cuoc LX, Scott RP, Bernardo MA, Chen DH, Valent B, Nelson RJ (1995) The relationship between lineage and virulence in *Pyricularia grisea* in the Philippines. Phytopathology 85:443–451

37. Saleh D, Milazzo J, Adreit H, Fournier E, Tharreau D (2014) South-East Asia is the center of origin, diversity and dispersion of the rice blast fungus, *Magnaporthe oryzae*. New Phytol 201:1440–1456

38. Saleh D, Xu P, Shen Y, Li C, Adreit H, Milazzo J, Ravigné V, Bazin E, Nottéghem J-L, Fournier E, Tharreau D (2012) Sex at the origin: an Asian population of the rice blast fungus *Magnaporthe oryzae* reproduces sexually. Mol Ecol 21:1330–1344

39. Orbach MJ, Chumley FG, Valent B (1996) Electrophoretic karyotypes of *Magnaporthe grisea* pathogens of diverse grasses. Mol Plant-Microbe Interact 9:261–271

40. Urashima AS, Igarashi S, Kato H (1993) Host range, mating type and fertility of *Pyricularia grisea* from wheat in Brazil. Plant Dis 77:1211–1216

41. Leung H, Borromeo ES, Bernardo MA, Notteghem JL (1988) Genetic analysis of virulence in the rice blast fungus *Magnaporthe grisea*. Phytopathology 78:1227–1233

42. Notteghem JL, Silue D (1992) Distribution of the mating type alleles in *Magnaporthe grisea* populations pathogenic on rice. Phytopathology 82:421–424

43. Chao CCT, Ellingboe AH (1991) Selection for mating competence in *Magnaporthe grisea* pathogenic to rice. Can J Bot 69:2130–2134

44. Dean RA, Talbot NJ, Ebbole DJ, Farman ML, Mitchell TK, Orbach MJ, Thon M, Kulkarni R, Xu J-R, Pan H, Read ND, Lee Y-H, Carbone I, Brown D, Oh YY, Donofrio N, Jeong JS, Soanes DM, Djonovic S, Kolomiets E, Renmeyer C, Li W, Harding M, Kim S, Lebrun MH, Bohnert H, Coughlan S, Butler J, Calvo S, Li-Jun M, Nicol R, Purcell S, Nusbaum C, Galagan JE, Birren BW (2005) The genome sequence of the rice blast fungus *Magnaporthe grisea*. Nature 434:980–986

45. Orbach MJ, Farrall L, Sweigard JA, Chumley FG, Valent B (2000) A telomeric avirulence gene AVR-Pita determines efficacy for the rice blast resistance gene pi-ta. Plant Cell 12:2019–2032

46. Sweigard JA, Carroll AM, Kang S, Farrall L, Chumley FG, Valent B (1995) Identification, cloning, and characterization of PWL2, a gene for host species specificity in the rice blast fungus. Plant Cell 7:1221–1233

47. Tosa Y, Inoue Y, Trinh TPV, Chuma I (2016) Genetic and molecular analyses of the incompatibility between Lolium isolates of *Pyricularia oryzae* and wheat. Physiol Mol Plant Pathol 95:84–86

48. Tosa Y, Tamba H, Tanaka K, Mayama S (2006) Genetic analysis of host species specificity of *Magnaporthe oryzae* isolates from rice and wheat. Phytopathology 96:480–484

49. Yaegashi H, Asaga K (1981) Further studies on the inheritance of pathogenicity in crosses of *Pyricularia oryzae* with *Pyricularia* sp. from finger millet. Ann Phytopathol Soc Japan 47:677–679

50. Chuma I, Shinogi T, Hosogi N, K-I I, Nakayashiki H, Park P, Tosa Y (2009) Cytological characteristics of microconidia of *Magnaporthe oryzae*. J Gen Plant Pathol 75:353–358

51. Kato H, Mayama S, Sekine R, Kanazawa E, Izutani Y, Urashima AS, Kunoh H (1994) Microconidium formation in *Magnaporthe grisea*. Ann Phytopathol Soc Japan 60:175–185

52. Zhang HL, Wu ZS, Wang CF, Li Y, Xu JR (2014) Germination and infectivity of microconidia in the rice blast fungus *Magnaporthe oryzae*. Nat Commun 5:4518

53. Zhang N, Cai G, Price DC, Crouch JA, Gladieux P, Hillman B, Khang CH, LeBrun M-H, Lee Y-H, Luo J, Qiu H, Veltri D, Wisecaver JH, Zhu J, Bhattacharya D (2018) Genome wide analysis of the transition to pathogenic lifestyles in Magnaporthales fungi. Sci Rep 8:5862

54. Sesma A, Osbourn AE (2004) The rice blast leaf pathogen undergoes developmental processes typical of root-infecting fungi. Nature 431:582–586

55. Marcel S, Sawers R, Oakeley E, Angliker H, Paszkowski U (2010) Tissue-adapted invasion strategies of the rice blast fungus *Magnaporthe oryzae*. Plant Cell 22:3177–3187

56. Crawford MS, Chumley FG, Weaver CG, Valent B (1986) Characterization of the heterokaryotic and vegetative diploid phases of *Magnaporthe grisea*. Genetics 114:1111–1129

57. Noguchi MT (2012) Virulence and aggressiveness of parasexual recombinants of the rice blast fungus *Magnaporthe oryzae*. Bulletin of the NARO Agricultural Research Center, pp 1–38
58. Noguchi MT, Yasuda N, Fujita Y (2006) Evidence of genetic exchange by parasexual recombination and genetic analysis of pathogenicity and mating type of parasexual recombinants in rice blast fungus, *Magnaporthe oryzae*. Phytopathology 96:746–750
59. Zeigler RS, Scott RP, Leung H, Bordeos AA, Kumar J, Nelson RJ (1997) Evidence of parasexual exchange of DNA in the rice blast fungus challenges its exclusive clonality. Phytopathology 87:284–294
60. Zeigler RS (1998) Recombination in *Magnaporthe grisea*. Annu Rev Phytopathol 36:249–275
61. Latterell FM, Rossi AE (1986) Longevity and pathogenic stability of *Pyricularia oryzae*. Phytopathology 76:231–235
62. Saleh D, Milazzo J, Adreit H, Tharreau D, Fournier E (2012) Asexual reproduction induces a rapid and permanent loss of sexual reproduction capacity in the rice fungal pathogen *Magnaporthe oryzae*: results of in vitro experimental evolution assays. BMC Evol Biol 12:42
63. Jia Y, Wamishe YA, Zhou B (2014) An expedited method for isolation of DNA for PCR from *Magnaporthe oryzae* stored on filter paper. Crop J 2:267–271
64. Chumley FG, Valent B (1990) Genetic analysis of melanin-deficient, nonpathogenic mutants of *Magnaporthe grisea*. Mol Plant-Microbe Interact 3:135–143
65. Thompson JE, Fahnestock S, Farrall L, Liao D-I, Valent B, Jordan DB (2000) The second naphthol reductase of fungal melanin biosynthesis in *Magnaporthe grisea*. J Biol Chem 275:34867–34872
66. Farman ML (2002) Meiotic deletion at the BUF1 locus of the fungus *Magnaporthe grisea* is controlled by interaction with the homologous chromosome. Genetics 160:137–148
67. Adachi K, Hamer JE (1998) Divergent cAMP signaling pathways regulate growth and pathogenesis in the rice blast fungus *Magnaporthe grisea*. Plant Cell 10:1361–1374
68. Hamer JE, Valent B, Chumley FG (1989) Mutations at the Smo genetic-locus affect the shape of diverse cell-types in the rice blast fungus. Genetics 122:351–361
69. Kershaw MJ, Basiewicz M, Soanes DM, Yan X, Ryder LS, Csukai M, Oses-Ruiz M, Valent B, Talbot NJ (2019) Conidial morphogenesis and septin-mediated plant infection require Smo1, a Ras GTPase-activating protein in *Magnaporthe oryzae*. Genetics 211:151–167
70. Farman ML (2007) Telomeres in the rice blast fungus *Magnaporthe oryzae*: the world of the end as we know it. FEMS Microbiol Lett 273:125–132
71. Rahnama M, Novikova O, Starnes JH, Zhang S, Chen L, Farman Mark L (2020) Transposon-mediated telomere destabilization: a driver of genome evolution in the blast fungus. Nucleic Acids Res 48:7197–7217
72. Ou SH (1985) Rice diseases, 2nd edn. Commonwealth Mycological Institute, Kew, UK
73. Ballini E, Morel J-B, Droc G, Price A, Courtois B, Notteghem J-L, Tharreau D (2008) A genome-wide meta-analysis of rice blast resistance genes and quantitative trait loci provides new insights into partial and complete resistance. Mol Plant-Microbe Interact 21:859–868
74. Wang BH, Ebbole DJ, Wang ZH (2017) The arms race between *Magnaporthe oryzae* and rice: diversity and interaction of Avr and R genes. J Integr Agric 16:2746–2760
75. Silué D, Notteghem JL, Tharreau D (1992) Evidence for a gene-for-gene relationship in the *Oryza sativa-Magnaporthe grisea* pathosystem. Phytopathology 82:577–580
76. Kroj T, Chanclud E, Michel-Romiti C, Grand X, Morel JB (2016) Integration of decoy domains derived from protein targets of pathogen effectors into plant immune receptors is widespread. New Phytol 210:618–626
77. Liu WD, Liu JL, Triplett L, Leach JE, Wang GL (2014) Novel insights into rice innate immunity against bacterial and fungal pathogens. Annu Rev Phytopathol 52:213–241
78. Yaegashi H (1978) Inheritance of pathogenicity in crosses of *Pyricularia* isolates from weeping love grass and finger millet. Ann Phytopathol Soc Japan 44:626–632
79. Kang S, Sweigard JA, Valent B (1995) The PWL host specificity gene family in the blast fungus *Magnaporthe grisea*. Mol Plant-Microbe Interact 8:939–948
80. Takabayashi N, Tosa Y, Oh HS, Mayama S (2002) A gene-for-gene relationship underlying the species-specific parasitism of Avena/Triticum isolates of *Magnaporthe grisea* on wheat cultivars. Phytopathology 92:1182–1188
81. Murakami J, Tosa Y, Kataoka T (2000) Analysis of host species specificity of *Magnaporthe*

*grisea* toward wheat using a genetic cross between isolates from wheat and foxtail millet. Phytopathology 90:1060–1067

82. Inoue Y, Vy TTP, Yoshida K, Asano H, Mitsuoka C, Asuke S, Anh VL, Cumagun CJR, Chuma I, Terauchi R, Kato K, Mitchell T, Valent B, Farman M, Tosa Y (2017) Evolution of the wheat blast fungus through functional losses in a host specificity determinant. Science 357:80–83
83. Bao J, Chen M, Zhong Z, Tang W, Lin L, Zhang X, Jiang H, Zhang D, Miao C, Tang H, Zhang J, Lu G, Ming R, Norvienyeku J, Wang B, Wang Z (2017) PacBio sequencing reveals transposable element as a key contributor to genomic plasticity and virulence variation in *Magnaporthe oryzae*. Mol Plant 10:1465–1468
84. Peng Z, Oliveira-Garcia E, Lin G, Hu Y, Dalby M, Migeon P, Tang H, Farman M, Cook D, White FF, Valent B, Liu S (2019) Effector gene reshuffling involves dispensable mini-chromosomes in the wheat blast fungus. PLoS Genet 15:e1008272
85. Luciano LBG, Tsai IJ, Chuma I, Tosa Y, Chen YH, Li JY, Li MY, Lu MYJ, Nakayashiki H, Li WH (2019) Blast fungal genomes show frequent chromosomal changes, gene gains and losses, and effector gene turnover. Mol Biol Evol 36:1148–1161
86. Chiapello H, Mallet L, Guerin C, Aguileta G, Amselem J, Kroj T, Ortega-Abboud E, Lebrun MH, Henrissat B, Gendrault A, Rodolphe F, Tharreau D, Fournier E (2015) Deciphering genome content and evolutionary relationships of isolates from the fungus *Magnaporthe oryzae* attacking different host plants. Genome Biol Evol 7:2896–2912
87. Yoshida K, Saunders DGO, Mitsuoka C, Natsume S, Kosugi S, Saitoh H, Inoue Y, Chuma I, Tosa Y, Cano LM, Kamoun S, Terauchi R (2016) Host specialization of the blast fungus *Magnaporthe oryzae* is associated with dynamic gain and loss of genes linked to transposable elements. BMC Genomics 17:370
88. Talbot NJ, Salch YP, Ma M, Hamer JE (1993) Karyotypic variability within clonal lineages of the rice blast fungus, *Magnaporthe grisea*. Appl Environ Microbiol 59:585–593
89. Raffaele S, Kamoun S (2012) Genome evolution in filamentous plant pathogens: why bigger can be better. Nat Rev Microbiol 10:417–430
90. Valent B, Khang CH (2010) Recent advances in rice blast effector research. Curr Opin Plant Biol 13:434–441
91. Khang CH, Park S-Y, Lee Y-H, Valent B, Kang S (2008) Genome organization and evolution of the AVR-Pita avirulence gene family in the *Magnaporthe grisea* species complex. Mol Plant-Microbe Interact 21:658–670
92. Huang J, Si W, Deng Q, Li P, Yang S (2014) Rapid evolution of avirulence genes in rice blast fungus *Magnaporthe oryzae*. BMC Genet 15:45
93. Wang X, Jia Y, Wamishe Y, Jia MH, Valent B (2017) Dynamic changes in the rice blast population in the USA over six decades. Mol Plant-Microbe Interact 30:803–812
94. Zhou E, Jia Y, Singh P, Correll JC, Lee FN (2007) Instability of the *Magnaporthe oryzae* avirulence gene AVR-Pita alters virulence. Fungal Genet Biol 44:1024–1034
95. Chuma I, Isobe C, Hotta Y, Ibaragi K, Futamata N, Kusaba M, Yoshida K, Terauchi R, Fujita Y, Nakayashiki H, Valent B, Tosa Y (2011) Multiple translocation of the AVR-Pita effector gene among chromosomes of the rice blast fungus *Magnaporthe oryzae* and related species. PLoS Pathog 7:e1002147
96. Ebbole DJ (2007) Magnaporthe as a model for understanding host-pathogen interactions. Annu Rev Phytopathol 45:437–456
97. Fernandez J, Wilson RA (2012) Why no feeding frenzy? Mechanisms of nutrient acquisition and utilization during infection by the rice blast fungus *Magnaporthe oryzae*. Mol Plant-Microbe Interact 25:1286–1293
98. Giraldo MC, Valent B (2013) Filamentous plant pathogen effectors in action. Nat Rev Microbiol 11:800–814
99. Talbot NJ (2003) On the trail of a cereal killer: exploring the biology of *Magnaporthe grisea*. Annu Rev Microbiol 57:177–202
100. Valent B (1990) Rice blast as a model system for plant pathology. Phytopathology 80:33–36
101. Valent B, Chumley FG (1991) Molecular genetic analysis of the rice blast fungus, *Magnaporthe grisea*. Annu Rev Phytopathol 29:443–467
102. Wilson RA, Talbot NJ (2009) Under pressure: investigating the biology of plant infection by *Magnaporthe oryzae*. Nat Rev Microbiol 7:185–195
103. Xu J-R (2000) MAP kinases in fungal pathogens. Fungal Genet Biol 31:137–152
104. Patkar RN, Naqvi NI (2017) Fungal manipulation of hormone-regulated plant defense. PLoS Pathog 13:e1006334

105. Liu W, Liu J, Ning Y, Ding B, Wang X, Wang Z, Wang G-L (2013) Recent progress in understanding PAMP- and effector-triggered immunity against the rice blast fungus *Magnaporthe oryzae*. Mol Plant 6 (3):605–620. https://doi.org/10.1093/mp/sst015

106. De la Concepcion JC, Franceschetti M, MacLean D, Terauchi R, Kamoun S, Banfield MJ (2019) Protein engineering expands the effector recognition profile of a rice NLR immune receptor. elife 8:e47713

107. Leach JE, Vera Cruz CM, Bai J, Leung H (2001) Pathogen fitness penalty as a predictor of durability of disease resistance genes. Annu Rev Phytopathol 39:187–224

108. Couch BC, Fudal I, Lebrun M-H, Tharreau D, Valent B, van Kim P, Notteghem J-L, Kohn LM (2005) Origins of host-specific populations of the blast pathogen *Magnaporthe oryzae* in crop domestication with subsequent expansion of pandemic clones on rice and weeds of rice. Genetics 170:613–630

109. Valent B, Singh PK, He X, Farman M, Tosa Y, Braun HJ (2020) CHAPTER 13: blast diseases: evolution and challenges of a staple food crop fungal pathogen. In: Ristaino JB, Records A (eds) Emerging plant diseases and global food security. American Phytopathological Society Press, St. Paul, MN, pp 267–292

# Chapter 2

# The Biology of Invasive Growth by the Rice Blast Fungus *Magnaporthe oryzae*

**Neftaly Cruz-Mireles, Iris Eisermann, Marisela Garduño-Rosales, Camilla Molinari, Lauren S. Ryder, Bozeng Tang, Xia Yan, and Nicholas J. Talbot**

## Abstract

This introductory chapter describes the life cycle of *Magnaporthe oryzae*, the causal agent of rice blast disease. During plant infection, *M. oryzae* forms a specialized infection structure called an appressorium, which generates enormous turgor, applied as a mechanical force to breach the rice cuticle. Appressoria form in response to physical cues from the hydrophobic rice leaf cuticle and nutrient availability. The signaling pathways involved in perception of surface signals are described and the mechanism by which appressoria function is also introduced. Re-polarization of the appressorium requires a septin complex to organize a toroidal F-actin network at the base of the cell. Septin aggregation requires a turgor-dependent sensor kinase, Sln1, necessary for re-polarization of the appressorium and development of a rigid penetration hypha to rupture the leaf cuticle. Once inside the plant, the fungus undergoes secretion of a large set of effector proteins, many of which are directed into plant cells using a specific secretory pathway. Here they suppress plant immunity, but can also be perceived by rice immune receptors, triggering resistances. *M. oryzae* then manipulates pit field sites, containing plasmodesmata, to facilitate rapid spread from cell to cell in plant tissue, leading to disease symptom development.

**Key words** Appressorium, Fungus, *Pyricularia*, Septins, Autophagy, Cell cycle, MAP kinase, Virulence, Plasmodesmata, Rice, Wheat, Pathogenesis

## 1 Introduction

The rice blast fungus has emerged as a major model system for understanding the molecular basis of fungal pathogenesis and the biology of plant immunity. Insights made from studying the fungus include a detailed understanding of the biology of infection structure development and the critical role of specific signal transduction pathways that regulate the initial stages of plant infection. These signaling functions are associated with broader physiological processes—such as starvation stress adaptation, surface perception, and autophagy—as well as fundamental cellular mechanisms involved in

Stefan Jacob (ed.), *Magnaporthe oryzae: Methods and Protocols*, Methods in Molecular Biology, vol. 2356, https://doi.org/10.1007/978-1-0716-1613-0_2, 

asymmetry generation, polarization, and cell cycle control. Definition of fungal effector proteins in the rice blast fungus has, meanwhile, led to new insight into the suppression of plant immunity and the specific recognition processes involved in immune receptor perception and disease resistance. When considered together, these discoveries have provided a platform for generating the most detailed understanding of a plant-fungal interaction. Such developments require the generation of new and specific methods, and this volume is significant in collecting the full spectrum of cell biological, biochemical, and physiological methods developed for studying this devastating pathogen. Here, we describe the life cycle of the fungus, highlighting what is understood about the biology of rice blast and where the methods described in the chapters that follow have been (or will be) pivotal to new understanding.

## 2 The Early Events of Plant Infection

### 2.1 The Life Cycle of Magnaporthe Oryzae

To cause infection, the rice blast fungus *Magnaporthe oryzae* (syn. *Pyricularia oryzae*) forms three-celled, teardrop-shaped (pyriform) conidia which develop under humid conditions, erupting at the tips of aerial conidiophores that emerge from disease lesions [1]. *M. oryzae* sporulates at dawn, at the maximum dew point, allowing dewdrop splash and wind to carry spores to neighboring plants and fields. This process is regulated by circadian clock control (*see* Chapter 13 by Griffin and Littlejohn). Conidia initially attach to the hydrophobic, waxy surface of a rice leaf using an adhesive called spore tip mucilage, released from an apical compartment in the spore [2]. In the presence of water, conidial germination occurs within 2 h and a narrow-polarized germ tube emerges from the apical cell of the conidium [3]. The germ tube tip begins to flatten at the leaf surface and swells in a process known as "hooking" [3]. The swollen tip of the germ tube then morphologically differentiates into a specialized single-celled infection structure called an "appressorium" (*see* Fig. 1a).

The dome-shaped appressorium is a prerequisite for plant infection and enables the fungus to physically rupture the plant host cuticle [1]. As the appressorium matures, it generates a melanin layer located between the appressorium plasma membrane and the cell wall [3] (for ultrastructural analysis of appressoria, *see* Chapter 6 by Rocha and Wilson). The melanin layer is porous, allowing water to be taken up by the appressorium while at the same time retaining larger molecules such as glycerol and other polyols to accumulate in the cell [4, 5]. When water enters the appressorium, against the high concentration gradient of glycerol, this generates hydrostatic turgor pressure that is rapidly translated into mechanical force to drive a narrow rigid penetration hypha from the base of the appressorium to rupture the rice leaf cuticle

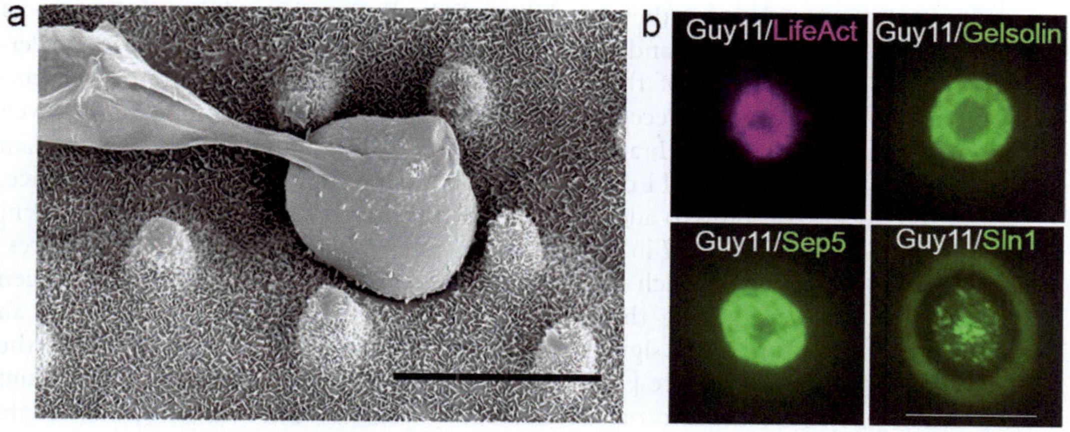


**Fig. 1** Appressorium morphogenesis in the rice blast fungus. (**a**) Scanning electron micrograph showing an appressorium on the rice leaf surface. The conidium has already collapsed due to autophagic cell death. Bar = 10 μm. Micrograph by Kim Findlay, John Innes Centre Bio-Imaging. (**b**) Reorganization of the fungal cytoskeleton at the appressorium pore. An F-actin toroidal network can be visualized by expression of LifeAct-mCherry and Gelsolin-GFP. This is scaffolded by a septin ring, shown here by expression of Sep5-GFP. The turgor sensor kinase Sln1-GFP localizes to the appressorium pore. Bar = 10 μm

[5, 6]. The fungus invaginates the host plant plasma membrane and rapidly fills the epidermal cells, secreting effectors to suppress the host immunity [7, 8]. Once the bulbous invasive hyphae fill the first infected host cell, the fungus accesses the neighboring plant cells through pit field sites where plasmodesmata accumulate [9]. Disease lesions then become visible 96 h after infection, and sporulation occurs under the conditions of high humidity. Conidia spread to neighboring plants and fields and the *M. oryzae* life cycle is complete [10, 11].

Although seldom observed in the field, *M. oryzae* also has the capacity to undergo sexual reproduction, in which two individuals of different mating type (Mat1-1 and Mat1-2) form a flask-shaped perithecium, which contains numerous asci, each of which carries eight ascospores—with each pair of spores representing one of the four products of meiosis. Sexual reproduction is common in some host-limited forms of *M. oryzae*, such as finger millet pathogens, for instance, but most rice-infecting strains of the fungus have lost fertility [10].

To initiate appressorium development, the fungus responds to physiological and biochemical cues, which include surface hardness, hydrophobicity, and rice leaf waxes [1]. These signals are translated through signaling pathways to trigger appressorium formation on an appropriate host cell surface [12]. The secreted class I hydrophobin gene *MPG1* is highly expressed during germination, appressorium formation, and plant infection [13]. The hydrophobin is secreted into the space between the fungal cell wall and plant tissue surface, self-assembling to form an amyloid structure which increases the wettability of the leaf surface, aiding the action of cutinases and triggering appressorium development [13, 14]. Perception of the hydrophobic cuticle involves integral G-protein-coupled receptors (GPCRs), such as Pth11, composed of seven transmembrane domains and an extracellular cysteine-rich domain [15]. Pth11 detects external physical cues from the rice leaf surface, leading to adenylate cyclase activation. Targeted gene replacement of *PTH11* in *M. oryzae* generates mutants unable to form appressoria, which are consequently non-pathogenic [16]. It has been suggested that the G-protein signal pathway also acts as an upstream signaling pathway for response to the hardness of the leaf surface [17]. The Gα subunit MagB suppresses appressorium formation on non-inductive surfaces and maintains adenylate cyclase in an inactive state [18]. The regulator of G-protein signaling proteins (RGSs) also controls MagB and cAMP levels during appressorium morphogenesis [19]. Rgs1 interacts with MagA, MagB, and MagC subunits in *M. oryzae* [1, 18, 20] and deletion of *RGS1* causes elevated cAMP levels [18], consistent with Rgs1 negatively regulating G-protein signaling through interaction with MagB and MagA. It is thought that upon activation from surface cues, these receptors transduce information to the protein kinase A-dependent cAMP response pathway and other signaling pathways [18, 21]. Cyclic AMP signaling transmits signals to downstream effectors that induce appressorium development [22–24]. *MAC1*, encoding adenylate cyclase, catalyzes production of cAMP from ATP, while cAMP phosphodiesterases hydrolyze cAMP [10, 19, 25]. Null mutants of *MAC1* cannot elaborate appressoria, but this phenotype can be remediated by exogenous cAMP [26]. Formation of appressoria on non-inductive hydrophilic surfaces can be induced by exogenous cAMP [27], and this ultimately activates the cAMP-dependent protein kinase A [28], a tetrameric holoenzyme which phosphorylates target proteins associated with appressorium maturation and turgor generation [24]. Cyclic AMP signaling is repressed by the target of rapamycin (TOR) [29] kinase and glutamine levels, regulated by the Asd4 transcription factor, which are key to the progression of appressorium development [30]. The precise control of sugar sensing and nitrogen source utilization are critical components of the regulation of appressorium function, which only occurs under starvation conditions [31, 32]. The analysis of cell signaling in *M. oryzae* has been greatly facilitated by methods to purify protein complexes, as described in Chapter 8 by Li and Wilson.

### 2.2 The Role of the Pmk1 MAP Kinase in Appressorium Development

The pathogenicity MAP kinase 1 (Pmk1) signaling pathway is a conserved signaling cascade in pathogenic fungi and has been reported to be crucial for plant infection in more than 20 plant pathogenic species [33, 34]. Many components of the pathway were identified initially by homology to the Fus3/Kss1 MAPK cascade in the model fungus *Saccharomyces cerevisiae* and, subsequently, by using *M. oryzae* as a guide.

In *M. oryzae*, it has long been known that the Pmk1 MAPK is required for appressorium morphogenesis. The absence of Pmk1 by generation of a Δ*pmk1* null mutant is sufficient to render the fungus non-pathogenic [23, 35], and Δ*pmk1* mutants fail to form appressoria, but are non-pathogenic even when spores are inoculated into wounded leaves. Recently, it has been shown that a conditionally inactive *pmk1*$^{AS}$ analogue-sensitive mutant of *M. oryzae* fails to move from rice cell-to-rice cell during invasive growth, implicating the MAPK in formation of specialized structures required to traverse the host cells [32]. A Δ*pmk1* mutant still responds to hydrophobic surfaces or exposure to exogenous cAMP by forming swollen germ tube tips [23]. The Pmk1 MAPK is expressed throughout vegetative growth of *M. oryzae*, and it has been reported that during appressorium maturation, GFP-Pmk1 localizes to the appressorium nucleus, consistent with the observations in other filamentous fungi that MAPKs can be translocated to the nucleus in response to specific stimuli [36].

During appressorium development, the Pmk1 MAP kinase is activated by its upstream MAPKK Mst7 and MAPKKK Mst11 [37]. Mst7 and Mst11 deletion mutants also fail to form appressoria and are non-pathogenic [37]. The Mst7 MAPKK appears to activate Pmk1 by phosphorylation of threonine and tyrosine residues in the well-conserved MAP kinase TXY motif [33, 37]. It has been reported that the interaction between Mst7 and Pmk1 occurs via a Mst7 MAPK-docking domain [38, 39]. Moreover, Mst7 forms homodimers and requires the action of thioredoxins to activate Pmk1 during ROS signaling [40]. The MAPKKK Mst11 can also be regulated by its self-inhibitory binding and through interaction with Ras proteins, via a Ras association domain [41, 42]. Both Mst11 and Mst7 form a complex with a scaffold protein, Mst50, that works as an adaptor [38, 43, 44]. Mst50 deletion mutants are defective in appressorium formation and non-pathogenic [44]. The Mst50 interaction with Mst7 and Mst11 is predicted to be direct, and the interaction between Mst50 and Mst11 occurs through its sterile α-motif (SAM) domain, although detailed in vivo binding assays are still necessary to confirm all of the Pmk1 signaling pathway interactions [44]. Ordering signaling pathways benefits from experimental methods such as bypass suppressor analysis, as described using the example of the osmosensing pathway (HOG pathway) by Jacob and Bersching in Chapter 18.

Upstream of the Mst50-Mst11-Mst7 complex, the G-beta subunit Mgb1 [45], the plasma membrane protein Pth11 [16, 46], and the sensing surface proteins Msb2 and Sho1 are necessary for initiation of appressorium morphogenesis [38]. Mgb1 is predicted to interact directly with Mst50, suggesting that Mst50 may integrate multiple upstream signals to activate the Pmk1 MAPK cascade [44]. Mgb1, Pth11, Msb2, and Sho1 also play a role in the surface recognition pathway and activation of cyclic AMP-protein kinase A (cAMP-PKA) [38]. There is undoubtedly a link between cAMP-PKA and Pmk1 pathways, but the exact mechanism by which this occurs remains unclear [47]. It is clear, however, that TOR kinase must remain inactive in order for appressorium development to proceed [29].

Little is known regarding the downstream targets of Pmk1, which regulate appressorium development. By yeast two-hybrid analysis, two Pmk1-interacting clones (PICs) were identified that serve roles in conidiation and appressorium differentiation [48]. PIC5 mutants are defective in germ tube growth and appressorium formation [48]. Transcription factors, such as the homeobox protein Hox7 and the C2/H2 Zn finger-domain protein Znf1, may operate downstream of the Pmk1 cascade and are also indispensable for appressorium formation [49, 50]. A comparative transcriptomics analysis of Δ*pmk1* mutants showed that 481 genes appear to be positively regulated by Pmk1 during germination, including known pathogenic components such as the GAS proteins and Pth11 [51]. Phosphoproteomic approaches will be necessary to understand the phosphorylation-dependent signal transduction events of the pathway, as detailed in Chapter 9 by Michna and Tenzer. These studies have benefited enormously from the ability to generate mutants very rapidly using targeted gene replacement methods and, more recently, the generation of CRISPR-Cas9-mediated genome editing, as described by Arazoe in Chapter 12.

### 2.3 Cell Cycle Control of Appressorium Development and Autophagic Cell Death

*M. oryzae* appressorium morphogenesis is tightly regulated by cell cycle progression [1, 52–54]. Cell cycle regulation, for example, is necessary to initiate the appearance of the fungal germ tube from the three-celled conidium. At this stage, one of the nuclei migrates into the germ tube, where it undergoes a single round of mitosis. The daughter nucleus then migrates to the tip of the germ tube, before isotropic swelling begins that leads to the formation of the appressorium. The other daughter nucleus returns to the conidium where it is degraded [1, 55]. Evidence implicating cell cycle control has been reported, for example, by generating temperature-sensitive mutations of the *NIMA* gene [52]. This gene is fundamental to mitosis and involved in the conserved DNA damage repair checkpoint pathway [52, 56]. When tested, together with a DNA synthesis inhibitor (hydroxyurea, HU), it was verified that successful execution of DNA replication (S-phase) is required for initiation of appressorium development [52]. Additionally, to ensure successful DNA duplication and segregation, cells arrest the cell cycle when DNA damage occurs [57]. This happens through inhibitory phosphorylation of the B-cyclin-CDK1 complex by two serine-threonine protein kinases of the DNA damage response (DDR) pathway [57]. Identification and mutant generation of the corresponding *M. oryzae* serine-threonine kinases involved revealed that they were both involved in the S-phase checkpoint during the early stages of the appressorium development [54]. However, this cell cycle checkpoint is not the only one that has been reported. There may also be a TOR-dependent G2 delay that is essential for appressorium autophagy to occur [29]. After the first round of mitosis, a threshold of turgor must also be reached in the appressorium, for its nucleus to transition from G1 to S-phase [53, 54]. This was tested using turgor-deficient mutants which are arrested in G1 and do not continue with infection [54]. This second cell cycle checkpoint appears to be controlled in a novel DDR-independent manner [53, 54]. These studies have benefited very significantly from live-cell imaging techniques, as detailed by Rogers et al. in Chapter 7.

Degradation of the remaining three conidial nuclei and collapse of the three-celled conidium (Fig. 1a) are also necessary for appressorium maturation and mediated by cell cycle control [1, 29, 53]. When mitosis is blocked, for example, appressorium formation is prevented, as well as conidial autophagy [1, 52]. Targeted deletion of *M. oryzae* autophagy-associated genes showed that the collapse of the three-celled conidium requires autophagy [1]. Two independent studies of the function of Atg1 and Atg8 demonstrated the requirement for autophagy in appressorium maturation [58, 59]. Autophagy mutants produce appressoria normally, but conidia fail to collapse, and, as a consequence, appressoria cannot re-polarize and are therefore not in a position to cause plant infection [4, 53, 58, 59]. Recent evidence has been presented that suggests that the final collapse and death of conidia requires a form of iron-dependent cell death, called ferroptosis, to occur [60]. Methods to study the onset of autophagy in *M. oryzae* are described by Li et al. in Chapter 14.

## 3 Appressorium Maturation

### 3.1 Turgor Generation and Turgor Sensing in M. oryzae

Appressoria in *M. oryzae* are simple unicellular structures, but they can be intricate multicellular structures, such as “infection cushions” in other pathogenic species [61–63]. Once mature the appressorium develops enormous turgor of up to 8.0 MPa, accumulating up to 3.0 M glycerol and other polyols to draw water into the cell by osmosis (Fig. 1a) [5]. The appressoria adhere tightly to the host surface aided by the action of mucilage, related to the spore tip mucilage that attaches conidia to the leaf surface, and the action of cutinases [64]. The appressorium develops a differentiated cell wall rich in melanin that is impermeable to glycerol which is also necessary for turgor generation. The melanin barrier allows water to enter the cell against the osmotic gradient and to retain glycerol, generating enormous hydrostatic pressure [2, 5]. Rusts and powdery mildews can form functional appressoria in the absence of water, suggesting the mechanism for turgor generation is distinct [47]. In fact, as we begin to evaluate the important criteria essential for successful appressorium turgor generation, it is apparent that not all appressoria generating fungi have the same requirements. Studies have, for example, revealed that appressorium turgor can be generated in the absence of melanin; in the soybean rust fungus *Phakopsora pachyrhizi*, for instance, non-melanized appressoria are still able to generate 5.13 MPa of pressure [65, 66]. Similarly, in the anthracnose pathogen of corn, *Colletotrichum graminicola*, turgor is still generated even when melanin biosynthesis is inhibited. Furthermore, cytorrhysis assay analysis of the osmolyte content using a method called Mach-Zehnder interferometry showed melanin is not required for solute accumulation and turgor generation [67]. This suggests that melanin does not provide the barrier for osmolytes in *C. graminicola* in the way it does in *M. oryzae* [47]. It is likely there are other components of the appressorium cell wall that can compensate for the function of melanin to achieve high turgor. A detailed biochemical and ultrastructural analysis of different appressoria cell walls is therefore vital in our understanding of how such extraordinary turgor pressures can be achieved.

Determining the precise mechanism for how the appressorium monitors turgor has been a long-standing conundrum. Previous work showed *M. oryzae* does not regulate appressorium turgor generation as cellular turgor is regulated in *S. cerevisiae*. In yeast, glycerol synthesis is regulated by the MAPK high-osmolarity glycerol response signaling pathway (HOG pathway), but in *M. oryzae*, the *HOG* homologue *OSM1* is dispensable for pathogenicity and glycerol production [68]. It is likely therefore that distinct biochemical routes for glycerol synthesis exist with different form of regulation. Conidia, for example, contain significant amounts of lipid, glycogen, trehalose, mannitol, and other storage products. Methods to study carbohydrate accumulation in *M. oryzae* are described by Grünewald et al., in Chapter 4. Considering conidia germinate in water droplets, glycerol must originate de novo from one or more of these sources (for a detailed review, *see* [6]).

Recent evidence has suggested that Sln1, a histidine-aspartate kinase, enables the appressorium to sense when a critical threshold of turgor has been reached, and thereby regulate host penetration [69]. Sln1 is an osmosensor in yeast that regulates hyperosmotic adaptation through the Hog1 MAPK pathway [70], and a previous study revealed *M. oryzae SLN1* to be necessary for pathogenicity [71]. A combination of cell biology, molecular genetics, discovery phosphoproteomics, and mathematical modeling demonstrated that the Sln1 sensor kinase responds to appressorium turgor by interacting with a set of upstream proteins that monitor cell expansion, including stretch-activated ion channel proteins, Mic1, Mic2, and Mic3, that maintain osmotic homeostasis within the cell. Once the required threshold of turgor is reached, it is maintained through the action of Sln1 by negatively regulating melanin biosynthesis and the cAMP/PKA pathway. Pkc1, the central regulator of the cell-integrity pathway, likely phosphorylates the phosphodiesterase PdeH to modulate cAMP levels and, as a consequence, will control lipolysis and glycerol production. Pkc1 may also target NoxR, a key regulator of the NADPH oxidase complex for regulated synthesis of ROS, which is necessary for septin-mediated cytoskeletal reorganization (Fig. 1b) [69, 72].

### 3.2 Cytoskeletal Remodeling During Appressorium Development

Experiments performed in the last decade began to explain the mechanism by which the re-establishment of the polarized growth occurs at the contact interface between the blast fungus and the plant host [73]. These morphogenetic phenomena ultimately lead to the generation and protrusion of the penetration peg into the host cuticle [73]. The penetration hypha emerges from the appressorium pore, which is the site of remodeling of the actin cytoskeleton [73, 74]. Filamentous actin (F-actin) is actively studied in filamentous fungi, due to its importance in polarized growth and morphogenesis [75–78]. F-actin interacts with associated proteins to confer stability to the different arrangements and organization of the cytoskeleton. F-actin can form high-order structures like patches, cables, and actomyosin contractile rings; these arrays have different functions and are responsible for distinct processes [75–77, 79–81]. A practical tool to visualize F-actin in live cells is the reporter LifeAct, which consists of the first 17 amino acids (MGVADLIKKFESISKEE) of the actin-binding protein Abp140 of budding yeast *Saccharomyces cerevisiae* [82]. Live-cell imaging of the actin cytoskeleton during appressorium development in *M. oryzae* has been carried out by expressing LifeAct-RFP (red fluorescent protein), which revealed an extensive toroidal F-actin network surrounding the appressorium pore at the base of the infection cell [73]. Assembly of the F-actin network during the generation of turgor in the appressorium indicates that specific reorientation of the cytoskeleton occurs at the base of the infection cell to achieve plant infection [73, 74]. It is here that septin GTPases play such an essential role by remodeling actin, forming a ring of 5.9 μm that colocalizes with the F-actin network at the appressorium pore [72, 73]. The septin ring also acts as a lateral diffusion barrier for specific actin-associated proteins, such as Las17, which polymerizes F-actin via the arp2/3 complex, and gelsolin, a key protein in the assembly and disassembly of the actin filaments [69, 73].

F-actin provides cortical rigidity prior to penetration peg emergence, as demonstrated in experiments performed in the *M. oryzae* mutants, Δ*cdc42* and Δ*chm1*, where the septin and F-actin networks are unable to form successfully [73]. Tea1, an ERM (ezrin, radixin, moesin) protein, tagged with GFP (green fluorescent protein), colocalizes with the F-actin and septin networks [73]. Interestingly, ERM proteins may link F-actin to the plasma membrane through an actin-binding domain in their C-terminus and an ERM domain that binds transmembrane proteins in their N-terminus [83], which supports the idea that ERM protein-actin linkages occur at the appressorium pore [73].

### 3.3 Septin-Dependent Plant Infection

Septins are guanosine triphosphatases (GTPases) [84] that were originally reported to play an essential role during cell division in *S. cerevisiae* [85] and are now recognized as key components of the cytoskeleton, like actin and microtubules. Septins are expressed in all eukaryotic organisms, with the exception of plants, and implicated in many cellular processes [28, 84, 86]. Septins consist of a highly conserved central GTP-binding domain, flanked by N- and C-terminal regions, and between the N-terminus and the GTP-binding domain is a short polybasic region (PBR). This region binds to negatively charged phospholipids and is responsible for interacting with membranes. A septin unique element [87] overlaps with the GTP-binding domain and is located toward the C-terminus. It is assumed that this region takes part in septin polymerization. The C-terminus itself is indispensable for septin-protein interactions [84, 88, 89]. Septins interact in a specific way with themselves and/or other septins at which they form dynamic hetero-oligomeric complexes [73, 90]. In *M. oryzae* the four core septin genes *SEP3*, *SEP4*, *SEP5*, and *SEP6* are homologues of the *S. cerevisiae* genes *CDC3*, *CDC10*, *CDC11*, and *CDC12*. During rice infection the hetero-oligomeric septin ring forms at the base of the appressorium (*see* Fig. 1b), and its formation and the dynamic recruitment of the F-actin have been visualized by quantitative 4D widefield fluorescence imaging [91]. The septin ring can only assemble after Sln1-dependent turgor sensing [69, 72] and progression of the appressorium nucleus through S-phase [54]. Sln1 acts through the Pkc1-dependent cell-integrity pathway to activate the Nox2-NoxR complex [69, 72] to regulate septin recruitment. The transcription factor Tpc1 regulates expression of the NADPH oxidase p22phox subunit Nox-D [92], and activation of the NADPH oxidase leads to regulated synthesis of ROS. The Ras-GTPase activating protein, Smo1, is also indispensable for septin recruitment and organization at the appressorium pore and negatively regulates the Ras2 signaling complex, while also directly interacting with the four core septins and components of the exocyst complex [93]. Additionally, assembly of the septin ring requires Cdc42 (a small GTPase of the Rho family) and Chm1 (a p21-activated kinase) which phosphorylates septins [73, 89]. The Mps1 MAPK and Mst12 are furthermore involved in septin assembly [94, 95]. In this context, sumoylation plays an additional role because attachment of a small ubiquitin-like modifier (SUMO) to septins is necessary for correct septin ring localization [96]. In this way septins recruit and organize the F-actin network at the appressorium pore, making phosphoinositide linkages between septins and the plasma membrane and facilitating F-actin plasma membrane linkages via ezrin, radixin, moesin (ERM) proteins, such as Tea1. The lateral diffusion barrier function of septins also secures the correct positioning of Las17 and Rvs167 [73], as well as the exocyst complex to the appressorium pore [97]. When considered together, septin-dependent functions are crucial for appressorium-mediated penetration to invade the leaf tissue.

## 4 Invasive Growth

### 4.1 BIC Formation and Translocation of Effectors

After successfully breaching the host plant cuticle, *M. oryzae* rapidly adapts to its new host environment [98, 99]. The fungus forms primary invasive hyphae after penetration of the rice epidermis [100]. Invasive hyphae develop as bulbous, branching structures that fill the epidermal cells [101–103]. Invasive hyphae are encased by the host extra-invasive hyphal membrane (EIHM), which provides a compartment to seal fungal invasive hyphae inside the host cell [101, 103]. During the biotrophic infection, the fungus secretes effector proteins and secondary metabolites into the host cells, which interfere with plant immunity response and physiology, thereby facilitating invasion and expansion [101, 102, 104]. This includes the modulation of host jasmonate signaling by a fungal monooxygenase to suppress immunity responses [105]. Methods to analyze secondary metabolism in *M. oryzae* are described in Chapter 3 by Skellam. Manipulation of the host environment includes suppressing the oxidative burst by oxidative denitrification of nitroalkanes using nitronate monooxygenase, encoded by the *NMO2* gene [106].

In the early stage of the infection, a structure is formed at the tip of the primary infecting hyphae, called the biotrophic interfacial complex (BIC) (*see* Fig. 2) [103, 104, 107]. Once the invasive hypha differentiates into bulbous branched structure, the BIC is positioned sub-apically and is a site where fungal effector proteins accumulate [108]. Formation of the BIC recurs when the fungus invades each new neighboring cell and requires *NMO2* [106]. Evidence has been reported which suggests that two distinct secretion pathways exist for the delivery of fungal effectors by *M. oryzae* [102]. Apoplastic effectors are secreted by a Golgi-dependent, brefeldin A-sensitive pathway, while effectors secreted into the host cells are secreted by a Golgi-independent, exocyst-dependent but brefeldin A-insensitive pathway. Cytoplasmic effectors also accumulate in the BIC, which appears to be the predominant

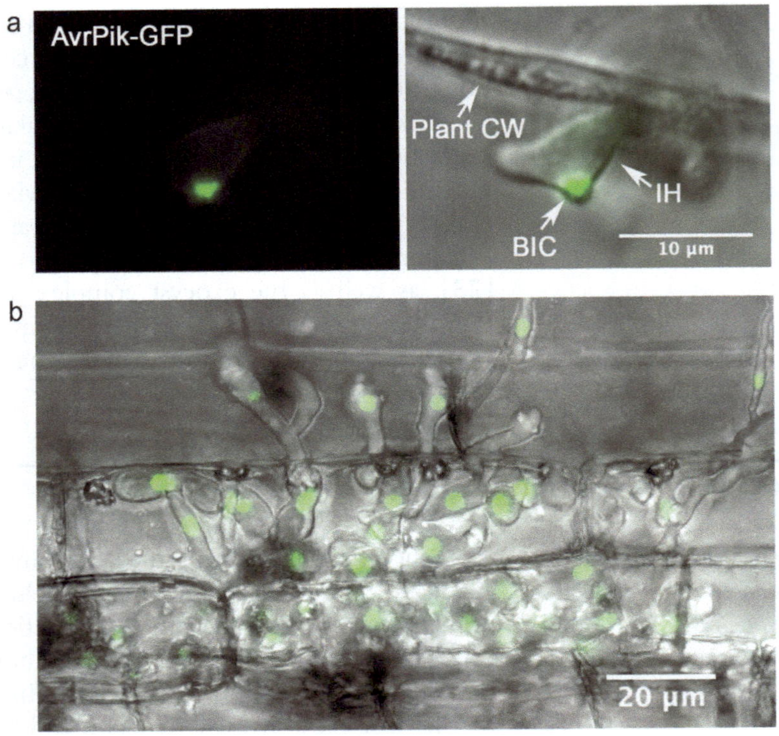


**Fig. 2** Laser confocal imaging of invasive growth of the rice blast fungus. (**a**) The biotrophic interfacial complex (BIC) formed in the first invaded cell, as visualized by the expression of AvrPik-GFP. Bar = 10 μm. (**b**) Invasive hyphae spreading throughout the rice tissue. Hyphae are expressing H1-GFP to show nuclei. Bar = 20 μm

route for their delivery, based on photobleaching experiments [102]. To obtain these detailed results was made possible through the methods to study the effector secretion systems of *M. oryzae* described by Oliveira-Garcia and Valent in Chapter 5. Recently, a gene involved in BIC formation, called *RBF1*, has been reported [108]. An *rbf1* mutant shows reduced virulence and triggers plant immunity in epidermal cells, consistent with the importance of the BIC in immune suppression mechanisms during invasive growth [108].

After 32–36 h *M. oryzae* invasive hyphae constrict and move through the pit fields to neighboring cells [35]. This process is regulated by the Pmk1 MAPK, which when conditionally inactivated prevents cell-to-cell spread by the fungus [32]. By contrast, the Mps1 MAPK is not required for cell-to-cell movement, but is involved in invasive hypha differentiation by coordinating the accumulation of α-1,3-glucans which may protect the fungus by avoiding degradation from plant-derived chitinases [94, 109]. The transcription factor Mig1, regulated by Mps1, is also necessary for invasive growth [110], while the Swi6 transcription factor is involved in cell wall integrity and the response to oxidative stress during invasive growth of *M. oryzae* [111]. It is also clear that the tight regulation of fungal metabolism is key to these morphogenetic transitions [112–114]. The maintenance of EIHM integrity, for example, requires *IMP1* and is tightly linked to the TOR-dependent regulation of autophagy, which must be tightly controlled during biotrophic growth [115].

Rice blast effectors have now been identified in significant numbers and their functions are beginning to be understood [116–118]. Methods to functionally analyze effectors are reported in Chapter 16 by Liu. Studies have shown diverse cytoplasmic effectors in *M. oryzae* accumulate BICs [104], including almost all the effector molecules known to be recognized by the immune receptor products of rice blast resistance genes, such as Avr-Pita, Avr-Pizt, Pwl1, Pwl2, and Bas1 [103], and the exocrine protein MC69 [103, 119, 120]. Pwl2 is not only secreted at the BIC, but has also been observed in adjacent uninfected rice cells, when these are plasmolyzed to concentrate the cytoplasmic signal [100]. Conversely, apoplastic effectors are secreted into the extracellular space between the hyphae and the EIHM, such as Bas4 and Slp1 [7, 104]. In the early stage of infection, Slp1, a LysM protein, aggregates at the interface between the plant and the fungus and can bind to chitin oligomers, released from the growing fungus, thereby inhibiting their perception by chitin receptors, such as CEBiP [7].

In many pathogens, effectors have been characterized based on their specific expression profiles in plants [121–125]. For instance, it has been shown that AvrE1 and HopM1 effectors secreted by phytopathogenic bacteria are essential bacterial proliferation in the apoplast of *Arabidopsis* [126]. In general, effectors are defined as virulence components which promote infection and manipulate plant immunity [127]. However, pathogen effectors can also be recognized by immune receptors [128–130]. In many species of plants and animals, intracellular receptors containing nucleotide-binding (NB) domains and leucine-rich repeat (LRR) proteins can detect effectors, either directly or indirectly [131, 132]. There are about 500 NB-LRRs known in rice, which are considered to act against *M. oryzae*, although only 33 discrete R-genes have been described in detail [133]. The first reported interaction was between Pi-ta and the fungal effector AVR-Pita, in which the rice immune receptor Pi-ta may bind to the AVR-Pita effector and lead to a hypersensitive reaction [134]. This also, however, allows the pathogen to conquer effector-triggered immunity by frequent mutation of effector-encoding genes that disrupt the specific interaction between effectors and the corresponding immune receptors in rice. The *Avr-Pik* alleles (*Pik/km/kp*) in *M. oryzae*, for example, show high levels of selection polymorphism [135, 136]. Similarly, the function of *Avr-CO39* has been abolished by transposon insertions in some isolates, leading to the gain of virulence in rice lines containing *CO39* [87]. Frequent mutations and transposon insertions therefore provide allelic diversity in avirulence genes in *M. oryzae*. For example, *Avr-Pita* is very unstable with a high level of translocation in the genome [137], and *Avr-Pi9* in *M. oryzae* has been subject to Mg-SINE insertion that has led to virulence on *Pi9*-containing rice cultivars [138]. A similar situation has occurred to drive the recent evolution of *M. oryzae* wheat blast isolates [139].

#### 4.2 Effector Function in M. oryzae

It is clear that *M. oryzae* secretes a very large battery of effector proteins into the plant cells during invasive growth to facilitate disease development, although the virulence functions of relatively few effectors have been determined [107, 140].

The majority of cytoplasmic effectors are small secreted proteins, with little homology outside of *M. oryzae*, although often well conserved among rice pathogens and other host-limited forms of the fungus [141]. Relatively few show clear homologies to proteins of known function; *AVR-Pita*, however, does encode a protein that contains a zinc-dependent metalloprotease domain, suggesting that it may cleave its cognate virulence target, although this has not been shown, and it is reported to bind directly to its cognate immune receptor Pi-ta [142]. The *ACE1* gene encodes a non-ribosomal peptide synthetase (NRPS)-polyketide synthase (PKS) hybrid enzyme. *ACE1* may therefore catalyze production of a secondary metabolite required for virulence that can be perceived by an immune receptor [143, 144]. The corresponding R gene *Pi33*, however, has not yet been identified [145]. A significant number of fungal effectors await characterization, including the family of fungal biotrophy associated secreted (BAS) effectors identified based on their expression profiles and localization during biotrophic growth [105]. Their function is still unknown because the generated mutants for these effectors failed to show clear phenotypes. However, further analysis has determined distinct cellular localizations—BAS1 shows accumulation at the BIC and cytoplasm, BAS2 and BAS3 localize to the BIC and the plant cell wall crossing points, and BAS4 accumulates in the outer space of the invasive hyphae, highlighting the potential diversity in their function.

Significant insight into the precise perception events leading to plant immunity has recently been gained using the rice blast pathosystem. Many NLR (nucleotide-binding, leucine-rich repeat) immune receptor proteins operate in pairs in a synergistic manner, where one NLR acts as a "sensor" and the second as a "helper" [146]. In *M. oryzae*, this organization is found, for example, in *Pii-1/Pii-2*, *Pik-1/Pik-2*, and *Pi-CO39/Pia* (Rga4/Rga5) gene pairs [147]. Recognition of the cognate Avr effector therefore requires a combination of two NLR proteins, which work together. Avr-Pia and Avr1-CO39 are both recognized by the RGA4/RGA5 pair of

NLRs, and this occurs by direct binding to a heavy metal-associated domain (HMA or RATX1), integrated into RGA5 [148]. This integrated domain may mimic the virulence target of these effectors because HMAs are implicated in host defense-associated ROS generation. The physical interaction of RGA4 and RGA5 prevents cell death that is mediated by RGA4, and the presence of AVR-Pia relieves that suppression, resulting in hypersensitive cell death [147]. The rice NLR, Pik-1, has also been shown to contain an integrated HMA domain, although inserted between the coiled domain and nucleotide-binding site (NB-ARC), and acts together with the helper NLR, Pik-2, in the recognition of AVR-Pik [135]. Multiple alleles of *Pik-1* can recognize different allelic forms of AVR-Pik. For example, *Pikp* rice lines are resistant to isolates carrying *AVR-Pik-D*, but susceptible to those expressing *AVR-Pik-E*, *AVR-Pik-A*, and *AVR-Pik-C* [135]. Structural analysis of the Pikp-HMA/AVR-PikD complex has revealed specific interactions leading to immunity [149], and this has led to further structure-function studies that have used protein engineering of the HMA domain of Pikp-1 to extend its recognition profile to different allelic variants of AVR-Pik [150, 151].

The concerted action of the battery of effectors secreted into the host cells leads to the suppression of plant defense and rapid proliferation of *M. oryzae* in rice tissue. Necrotic lesions then form, and host cell death is elicited. Conidiation occurs from disease lesions and the fungus spreads to new host plants. Methods to detect fungal colonization of seeds are described by Chadha in Chapter 15, and methods to identify novel fungicides for rice blast control are reported by Bohnert et al., in Chapter 10.

In summary, the life cycle of *M. oryzae* is exemplified by specific, highly orchestrated developmental transitions and precise adaptations to biotrophic growth in plant tissue, such as the evolution of a large family of effector proteins that can suppress immunity and overwhelm plant defenses, allowing the fungus to grow very rapidly in host tissue and cause blast disease. It is for these reasons that the rice blast fungus constitutes such a difficult pathogen to control. The methods in this volume do, however, provide the tools to make very rapid progress in understanding rice blast disease, which will be pivotal to its durable control.

## References


1. Wilson RA, Talbot NJ (2009) Under pressure: investigating the biology of plant infection by Magnaporthe oryzae. Nat Rev Microbiol 7:185–195
2. Hamer JE, Howard RJ, Chumley FG, Valent B (1988) A mechanism for surface attachment in spores of a plant pathogenic fungus. Science 239:288
3. Bourett TM, Howard RJ (1990) In vitro development of penetration structures in the rice blast fungus Magnaporthe grisea. Can J Bot 68:329–342

4. Kershaw MJ, Talbot NJ (2009) Genome-wide functional analysis reveals that infection-associated fungal autophagy is necessary for rice blast disease. Proc Natl Acad Sci U S A 106:15967–15972
5. de Jong JC, McCormack BJ, Smirnoff N, Talbot NJ (1997) Glycerol generates turgor in rice blast. Nature 389:244–244
6. Foster AJ, Ryder LS, Kershaw MJ, Talbot NJ (2017) The role of glycerol in the pathogenic lifestyle of the rice blast fungus Magnaporthe oryzae. Environ Microbiol 19(3):1008–1016
7. Mentlak TA, Kombrink A, Shinya T, Ryder LS, Otomo I, Saitoh H, Terauchi R, Nishizawa Y, Shibuya N, Thomma BP, Talbot NJ (2012) Effector-mediated suppression of chitin-triggered immunity by Magnaporthe oryzae is necessary for rice blast disease. Plant Cell 24:322–335
8. Yan X, Talbot NJ (2016) Investigating the cell biology of plant infection by the rice blast fungus Magnaporthe oryzae. Curr Opin Microbiol 34:147–153
9. Martin-Urdiroz M, Oses-Ruiz M, Ryder LS, Talbot NJ (2016) Investigating the biology of plant infection by the rice blast fungus Magnaporthe oryzae. Fungal Genet Biol 90:61–68
10. Talbot NJ (2003) On the trail of a cereal killer: exploring the biology of Magnaporthe grisea. Ann Rev Microbiol 57:177–202
11. Wang X, Valent B (2009) Advances in genetics, genomics and control of rice blast disease. Springer Science & Business Media, Berlin/Heidelberg
12. Liu W, Zhou X, Li G, Li L, Kong L, Wang C, Zhang H, Xu J-R (2011) Multiple plant surface signals are sensed by different mechanisms in the rice blast fungus for appressorium formation. PLoS Pathog 7:e1001261
13. Talbot NJ, Ebbole DJ, Hamer JE (1993) Identification and characterization of MPG1, a gene involved in pathogenicity from the rice blast fungus Magnaporthe grisea. Plant Cell 5:1575–1590
14. Pham CL, Rey A, Lo V, Soules M, Ren Q, Meisl G, Knowles TP, Kwan AH, Sunde M (2016) Self-assembly of MPG1, a hydrophobin protein from the rice blast fungus that forms functional amyloid coatings, occurs by a surface-driven mechanism. Sci Rep 6:25288
15. Kulkarni RD, Thon MR, Pan H, Dean RA (2005) Novel G-protein-coupled receptor-like proteins in the plant pathogenic fungus Magnaporthe grisea. Genome Biol 6:R24
16. DeZwaan TM, Carroll AM, Valent B, Sweigard JA (1999) Magnaporthe grisea pth11p is a novel plasma membrane protein that mediates appressorium differentiation in response to inductive substrate cues. Plant Cell 11:2013–2030
17. Liu S, Dean RA (1997) G protein α subunit genes control growth, development, and pathogenicity of Magnaporthe grisea. Mol Plant-Microbe Interact 10:1075–1086
18. Liu H, Suresh A, Willard FS, Siderovski DP, Lu S, Naqvi NI (2007) Rgs1 regulates multiple Gα subunits in Magnaporthe pathogenesis, asexual growth and thigmotropism. EMBO J 26:690–700
19. Fang EG, Dean RA (2000) Site-directed mutagenesis of the magB gene affects growth and development in Magnaporthe grisea. Mol Plant-Microbe Interact 13:1214–1227
20. Ramanujam R, Yishi X, Liu H, Naqvi NI (2012) Structure-function analysis of Rgs1 in Magnaporthe oryzae: role of DEP domains in subcellular targeting. PLoS One 7:e41084
21. Bölker M (1998) Sex and crime: heterotrimeric G proteins in fungal mating and pathogenesis. Fungal Genet Biol 25:143–156
22. Kang SH, Khang CH, Lee Y-H (1999) Regulation of cAMP-dependent protein kinase during appressorium formation in Magnaporthe grisea. FEMS Microbiol Lett 170:419–423
23. Xu J-R, Hamer JE (1996) MAP kinase and cAMP signaling regulate infection structure formation and pathogenic growth in the rice blast fungus Magnaporthe grisea. Genes Dev 10:2696–2706
24. Adachi K, Hamer JE (1998) Divergent cAMP signaling pathways regulate growth and pathogenesis in the rice blast fungus Magnaporthe grisea. Plant Cell 10:1361–1373
25. Dean RA, Talbot NJ, Ebbole DJ, Farman ML, Mitchell TK, Orbach MJ, Thon M, Kulkarni R, Xu J-R, Pan H, Read ND, Lee Y-H, Carbone I, Brown D, Oh YY, Donofrio N, Jeong JS, Soanes DM, Djonovic S, Kolomiets E, Rehmeyer C, Li W, Harding M, Kim S, Lebrun M-H, Bohnert H, Coughlan S, Butler J, Calvo S, Ma L-J, Nicol R, Purcell S, Nusbaum C, Galagan JE, Birren BW (2005) The genome sequence of the rice blast fungus Magnaporthe grisea. Nature 434:980–986
26. Choi W, Dean RA (1997) The adenylate cyclase gene MAC1 of Magnaporthe grisea controls appressorium formation and other aspects of growth and development. Plant Cell 9:1973–1983
27. Gilbert RD, Johnson AM, Dean RA (1996) Chemical signals responsible for appressorium

formation in the rice blast fungus Magnaporthe grisea. Physiol Mol Plant Pathol 48:335–346

28. Douglas LM, Alvarez FJ, McCreary C, Konopka JB (2005) Septin function in yeast model systems and pathogenic fungi. Eukaryot Cell 4:1503–1512
29. Dean R, Van Kan JA, Pretorius ZA, Hammond-Kosack KE, Di Pietro A, Spanu PD, Rudd JJ, Dickman M, Kahmann R, Ellis J, Foster GD (2012) The top 10 fungal pathogens in molecular plant pathology. Mol Plant Pathol 13:414–430
30. Marroquin-Guzman M, Wilson RA (2015) GATA-dependent glutaminolysis drives appressorium formation in Magnaporthe oryzae by suppressing TOR inhibition of cAMP/PKA signaling. PLoS Pathog 11:e1004851
31. Wilson RA, Gibson RP, Quispe CF, Littlechild JA, Talbot NJ (2010) An NADPH-dependent genetic switch regulates plant infection by the rice blast fungus. Proc Natl Acad Sci U S A 107:21902–21907
32. Fernandez J, Wright JD, Hartline D, Quispe CF, Madayiputhiya N, Wilson RA (2012) Principles of carbon catabolite repression in the rice blast fungus: Tps1, Nmr1-3, and a MATE-family pump regulate glucose metabolism during infection. PLoS Genet 8: e1002673
33. Jiang C, Zhang X, Liu H, Xu J-R (2018) Mitogen-activated protein kinase signaling in plant pathogenic fungi. PLoS Pathog 14: e1006875
34. Turrà D, Segorbe D, Di Pietro A (2014) Protein kinases in plant-pathogenic fungi: conserved regulators of infection. Annu Rev Phytopathol 52:267–288
35. Sakulkoo W, Osés-Ruiz M, Oliveira Garcia E, Soanes DM, Littlejohn GR, Hacker C, Correia A, Valent B, Talbot NJ (2018) A single fungal MAP kinase controls plant cell-to-cell invasion by the rice blast fungus. Science 359:1399–1403
36. Bruno KS, Tenjo F, Li L, Hamer JE, Xu JR (2004) Cellular localization and role of kinase activity of PMK1 in Magnaporthe grisea. Eukaryot Cell 3:1525–1532
37. Zhao X, Kim Y, Park G, Xu J-R (2005) A mitogen-activated protein kinase Cascade regulating infection-related morphogenesis in Magnaporthe grisea. Plant Cell Online 17:1317–1329
38. Li G, Zhou X, Xu J-R (2012) Genetic control of infection-related development in Magnaporthe oryzae. Curr Opin Microbiol 15:678–684
39. Zhao X, Xu J-R (2007) A highly conserved MAPK-docking site in Mst7 is essential for Pmk1 activation in Magnaporthe grisea. Mol Microbiol 63:881–894
40. Zhang S, Jiang C, Zhang Q, Qi L, Li C, Xu J-R (2016) Thioredoxins are involved in the activation of the PMK1 MAP kinase pathway during appressorium penetration and invasive growth in Magnaporthe oryzae. Environ Microbiol 18:3768–3784
41. Qi L, Kim Y, Jiang C, Li Y, Peng Y, Xu J-R (2015) Activation of Mst11 and feedback inhibition of germ tube growth in Magnaporthe oryzae. Mol Plant-Microbe Interact 28:881–891
42. Zhou X, Zhao X, Xue C, Dai Y, Xu J-R (2014) Bypassing both surface attachment and surface recognition requirements for Appressorium formation by overactive Ras signaling in Magnaporthe oryzae. Mol Plant-Microbe Interact 27:996–1004
43. Li G, Zhang X, Tian H, Choi Y-E, Tao WA, Xu J-R (2017) MST50 is involved in multiple MAP kinase signaling pathways in Magnaporthe oryzae. Environ Microbiol 19:1959–1974
44. Park G, Xue C, Zhao X, Kim Y, Orbach M, Xu JR (2006) Multiple upstream signals converge on the adaptor protein Mst50 in Magnaporthe grisea. Plant Cell 18:2822–2835
45. Nishimura M, Park G, Xu J-R (2003) The G-beta subunit MGB1 is involved in regulating multiple steps of infection-related morphogenesis in Magnaporthe grisea. Mol Microbiol 50:231–243
46. Kou Y, Tan YH, Ramanujam R, Naqvi NI (2017) Structure-function analyses of the Pth11 receptor reveal an important role for CFEM motif and redox regulation in rice blast. New Phytol 214:330–342
47. Ryder LS, Talbot NJ (2015) Regulation of appressorium development in pathogenic fungi. Curr Opin Plant Biol 26:8–13
48. Zhang H, Xue C, Kong L, Li G, Xu J-R (2011) A Pmk1-interacting gene is involved in appressorium differentiation and plant infection in Magnaporthe oryzae. Eukaryot Cell 10:1062–1070
49. Kim S, Park SY, Kim KS, Rho HS, Chi MH, Choi J, Park J, Kong S, Park J, Goh J, Lee YH (2009) Homeobox transcription factors are required for conidiation and appressorium development in the rice blast fungus Magnaporthe oryzae. PLoS Genet 5:e1000757
50. Yue X, Que Y, Xu L, Deng S, Peng Y, Talbot NJ, Wang Z (2016) ZNF1 encodes a putative C2H2 zinc-finger protein essential for

appressorium differentiation by the Rice blast fungus Magnaporthe oryzae. Mol Plant-Microbe Interact 29:22–35

51. Soanes DM, Chakrabarti A, Paszkiewicz KH, Dawe AL, Talbot NJ (2012) Genome-wide transcriptional profiling of appressorium development by the Rice blast fungus Magnaporthe oryzae. PLoS Pathog 8:e1002514

52. Saunders DG, Aves SJ, Talbot NJ (2010) Cell cycle-mediated regulation of plant infection by the rice blast fungus. Plant Cell 22:497–507

53. Oses-Ruiz M, Talbot NJ (2017) Cell cycle-dependent regulation of plant infection by the rice blast fungus Magnaporthe oryzae. Commun Integr Biol 10:e1372067

54. Oses-Ruiz M, Sakulkoo W, Littlejohn GR, Martin-Urdiroz M, Talbot NJ (2017) Two independent S-phase checkpoints regulate appressorium-mediated plant infection by the rice blast fungus Magnaporthe oryzae. Proc Natl Acad Sci U S A 114:E237–E244

55. Saunders DG, Dagdas YF, Talbot NJ (2010) Spatial uncoupling of mitosis and cytokinesis during appressorium-mediated plant infection by the rice blast fungus Magnaporthe oryzae. Plant Cell 22:2417–2428

56. Osmani AH, O'Donnell K, Pu RT, Osmani SA (1991) Activation of the nimA protein kinase plays a unique role during mitosis that cannot be bypassed by absence of the bimE checkpoint. EMBO J 10:2669–2679

57. Rhind N, Russell P (1998) Mitotic DNA damage and replication checkpoints in yeast. Curr Opin Cell Biol 10:749–758

58. Veneault-Fourrey C, Barooah M, Egan M, Wakley G, Talbot NJ (2006) Autophagic fungal cell death is necessary for infection by the rice blast fungus. Science 312:580–583

59. Liu XH, Lu JP, Zhang L, Dong B, Min H, Lin FC (2007) Involvement of a Magnaporthe grisea serine/threonine kinase gene, MgATG1, in appressorium turgor and pathogenesis. Eukaryot Cell 6:997–1005

60. Shen Q, Liang M, Yang F, Deng YZ, Naqvi NI (2020) Ferroptosis contributes to developmental cell death in rice blast. New Phytol 227(6):1831–1846

61. Armentrout V, Downer A (1987) Infection cushion development by Rhizoctonia solani on cotton. Phytopathology 77:619–623

62. Mendgen K, Hahn M, Deising H (1996) Morphogenesis and mechanisms of penetration by plant pathogenic fungi. Annu Rev Phytopathol 34:367–386

63. Talbot NJ (2019) Appressoria. Curr Biol 29: R144–r146

64. Skamnioti P, Gurr SJ (2007) Magnaporthe grisea Cutinase2 mediates appressorium differentiation and host penetration and is required for full virulence. Plant Cell 19:2674

65. Chang HX, Miller LA, Hartman GL (2014) Melanin-independent accumulation of turgor pressure in appressoria of Phakopsora pachyrhizi. Phytopathology 104:977–984

66. Loehrer M, Botterweck J, Jahnke J, Mahlmann DM, Gaetgens J, Oldiges M, Horbach R, Deising H, Schaffrath U (2014) In vivo assessment by Mach–Zehnder double-beam interferometry of the invasive force exerted by the Asian soybean rust fungus (Phakopsora pachyrhizi). New Phytol 203:620–631

67. Ludwig N, Löhrer M, Hempel M, Mathea S, Schliebner I, Menzel M, Kiesow A, Schaffrath U, Deising HB, Horbach R (2013) Melanin is not required for turgor generation but enhances cell-wall rigidity in appressoria of the corn pathogen Colletotrichum graminicola. Mol Plant-Microbe Interact 27:315–327

68. Dixon KP, Xu JR, Smirnoff N, Talbot NJ (1999) Independent signaling pathways regulate cellular turgor during hyperosmotic stress and appressorium-mediated plant infection by Magnaporthe grisea. Plant Cell 11:2045

69. Ryder LS, Dagdas YF, Kershaw MJ, Venkataraman C, Madzvamuse A, Yan X, Cruz-Mireles N, Soanes DM, Oses-Ruiz M, Styles V, Sklenar J, Menke FLH, Talbot NJ (2019) A sensor kinase controls turgor-driven plant infection by the rice blast fungus. Nature 574:423–427

70. Tao W, Deschenes RJ, Fassler JS (1999) Intracellular glycerol levels modulate the activity of Sln1p, a *Saccharomyces cerevisiae* two-component regulator. J Biol Chem 274:360–367

71. Zhang H, Liu K, Zhang X, Song W, Zhao Q, Dong Y, Guo M, Zheng X, Zhang Z (2010) A two-component histidine kinase, MoSLN1, is required for cell wall integrity and pathogenicity of the rice blast fungus, Magnaporthe oryzae. Curr Genet 56(6):517–528

72. Ryder LS, Dagdas YF, Mentlak TA, Kershaw MJ, Thornton CR, Schuster M, Chen J, Wang Z, Talbot NJ (2013) NADPH oxidases regulate septin-mediated cytoskeletal remodeling during plant infection by the rice blast fungus. Proc Natl Acad Sci 110:3179

73. Dagdas YF, Yoshino K, Dagdas G, Ryder LS, Bielska E, Steinberg G, Talbot NJ (2012) Septin-mediated plant cell invasion by the rice blast fungus, Magnaporthe oryzae. Science 336:1590–1595

74. Bourett TM, Howard RJ (1992) Actin in penetration pegs of the fungal rice blast pathogen, Magnaporthe grisea. Protoplasma 168:20–26
75. Delgado-Álvarez DL, Callejas-Negrete OA, Gomez N, Freitag M, Roberson RW, Smith LG, Mouriño-Pérez RR (2010) Visualization of F-actin localization and dynamics with live cell markers in Neurospora crassa. Fungal Genet Biol 47:573–586
76. Delgado-Álvarez DL, Bartnicki-García S, Seiler S, Mouriño-Pérez RR (2014) Septum development in Neurospora crassa: the septal actomyosin tangle. PLoS One 9:e96744
77. Upadhyay S, Shaw BD (2008) The role of actin, fimbrin and endocytosis in growth of hyphae in aspergillus nidulans. Mol Microbiol 68:690–705
78. Berepiki A, Lichius A, Shoji JY, Tilsner J, Read ND (2010) F-actin dynamics in Neurospora crassa. Eukaryot Cell 9:547–557
79. Roberson RW (1992) The actin cytoskeleton in hyphal cells of Sclerotium rolfsii. Mycologia 84:41–51
80. Taheri-Talesh N, Horio T, Araujo-Bazán L, Dou X, Espeso EA, Peñalva MA, Osmani SA, Oakley BR (2008) The tip growth apparatus of aspergillus nidulans. Mol Biol Cell 19:1439–1449
81. Berepiki A, Lichius A, Read ND (2011) Actin organization and dynamics in filamentous fungi. Nat Rev Microbiol 9:876–887
82. Riedl J, Crevenna AH, Kessenbrock K, Yu JH, Neukirchen D, Bista M, Bradke F, Jenne D, Holak TA, Werb Z, Sixt M, Wedlich-Soldner R (2008) Lifeact: a versatile marker to visualize F-actin. Nat Methods 5:605–607
83. Gilden J, Krummel MF (2010) Control of cortical rigidity by the cytoskeleton: emerging roles for septins. Cytoskeleton (Hoboken) 67:477–486. https://doi.org/10.1002/cm.20461
84. Van Ngo H, Mostowy S (2019) Role of septins in microbial infection. J Cell Sci 132: jcs226266. https://journals.biologists.com/jcs/article/132/9/jcs226266/57414/Role-of-septins-in-microbial-infection
85. Hartwell LH (1971) Genetic control of the cell division cycle in yeast IV. Genes controlling bud emergence and cytokinesis. Exp Cell Res 69:265–276
86. Spiliotis ET, Nelson WJ (2006) Here come the septins: novel polymers that coordinate intracellular functions and organization. J Cell Sci 119:4–10
87. Tosa Y, Osue J, Eto Y, Oh H-S, Nakayashiki H, Mayama S, Leong SA (2005) Evolution of an Avirulence gene, AVR1-CO39, concomitant with the evolution and differentiation of Magnaporthe oryzae. Mol Plant-Microbe Interact 18:1148–1160
88. Sirajuddin M, Farkasovsky M, Zent E, Wittinghofer A (2009) GTP-induced conformational changes in septins and implications for function. Proc Natl Acad Sci U S A 106:16592–16597
89. Versele M, Thorner J (2004) Septin collar formation in budding yeast requires GTP binding and direct phosphorylation by the PAK, Cla4. J Cell Biol 164:701–715
90. Sirajuddin M, Farkasovsky M, Hauer F, Kuhlmann D, Macara IG, Weyand M, Stark H, Wittinghofer A (2007) Structural insight into filament formation by mammalian septins. Nature 449:311–315
91. Dulal N, Rogers A, Wang Y, Egan M (2020) Dynamic assembly of a higher-order septin structure during appressorium morphogenesis by the rice blast fungus. Fungal Genet Biol 140:103385
92. Galhano R, Illana A, Ryder LS, Rodriguez-Romero J, Demuez M, Badaruddin M, Martinez-Rocha AL, Soanes DM, Studholme DJ, Talbot NJ, Sesma A (2017) Tpc1 is an important Zn(II)2Cys6 transcriptional regulator required for polarized growth and virulence in the rice blast fungus. PLoS Pathog 13:e1006516
93. Kershaw MJ, Basiewicz M, Soanes DM, Yan X, Ryder LS, Csukai M, Oses-Ruiz M, Valent B, Talbot NJ (2019) Conidial morphogenesis and Septin-mediated plant infection require Smo1, a Ras GTPase-activating protein in Magnaporthe oryzae. Genetics 211:151–167
94. Xu JR, Staiger CJ, Hamer JE (1998) Inactivation of the mitogen-activated protein kinase Mps1 from the rice blast fungus prevents penetration of host cells but allows activation of plant defense responses. Proc Natl Acad Sci U S A 95:12713–12718
95. Park G, Xue C, Zheng L, Lam S, Xu JR (2002) MST12 regulates infectious growth but not appressorium formation in the rice blast fungus Magnaporthe grisea. Mol Plant-Microbe Interact 15:183–192
96. Liu C, Li Z, Xing J, Yang J, Wang Z, Zhang H, Chen D, Peng YL, Chen XL (2018) Global analysis of sumoylation function reveals novel insights into development

and appressorium-mediated infection of the rice blast fungus. New Phytol 219:1031–1047

97. Gupta YK, Dagdas YF, Martinez-Rocha A-L, Kershaw MJ, Littlejohn GR, Ryder LS, Sklenar J, Menke F, Talbot NJ (2015) Septin-dependent assembly of the exocyst is essential for plant infection by Magnaporthe oryzae. Plant Cell 27:3277

98. Armijo G, Schlechter R, Agurto M, Muñoz D, Nuñez C, Arce-Johnson P (2016) Grapevine pathogenic microorganisms: understanding infection strategies and host response scenarios. Front Plant Sci 7:382

99. Fernandez J, Orth K (2018) Rise of a cereal killer: the biology of Magnaporthe oryzae biotrophic growth. Trends Microbiol 26:582–597

100. Yi M, Valent B (2013) Communication between filamentous pathogens and plants at the biotrophic interface. Annu Rev Phytopathol 51:587–611

101. Kankanala P, Czymmek K, Valent B (2007) Roles for rice membrane dynamics and plasmodesmata during biotrophic invasion by the blast fungus. Plant Cell 19:706–724

102. Giraldo MC, Dagdas YF, Gupta YK, Mentlak TA, Yi M, Martinez-Rocha AL, Saitoh H, Terauchi R, Talbot NJ, Valent B (2013) Two distinct secretion systems facilitate tissue invasion by the rice blast fungus Magnaporthe oryzae. Nat Commun 4:1996

103. Khang CH, Berruyer R, Giraldo MC, Kankanala P, Park SY, Czymmek K, Kang S, Valent B (2010) Translocation of Magnaporthe oryzae effectors into rice cells and their subsequent cell-to-cell movement. Plant Cell 22:1388–1403

104. Mosquera G, Giraldo MC, Khang CH, Coughlan S, Valent B (2009) Interaction transcriptome analysis identifies Magnaporthe oryzae BAS1-4 as biotrophy-associated secreted proteins in rice blast disease. Plant Cell 21:1273–1290

105. Patkar RN, Benke PI, Qu Z, Chen YY, Yang F, Swarup S, Naqvi NI (2015) A fungal monooxygenase-derived jasmonate attenuates host innate immunity. Nat Chem Biol 11:733–740

106. Marroquin-Guzman M, Hartline D, Wright JD, Elowsky C, Bourret TJ, Wilson RA (2017) The Magnaporthe oryzae nitrooxidative stress response suppresses rice innate immunity during blast disease. Nat Microbiol 2:17054

107. Valent B, Khang CH (2010) Recent advances in rice blast effector research. Curr Opin Plant Biol 13:434–441

108. Nishimura T, Mochizuki S, Ishii-Minami N, Fujisawa Y, Kawahara Y, Yoshida Y, Okada K, Ando S, Matsumura H, Terauchi R, Minami E, Nishizawa Y (2016) Magnaporthe oryzae glycine-rich secretion protein, Rbf1 critically participates in pathogenicity through the focal formation of the biotrophic interfacial complex. PLoS Pathog 12:e1005921

109. Fujikawa T, Kuga Y, Yano S, Yoshimi A, Tachiki T, Abe K, Nishimura M (2009) Dynamics of cell wall components of Magnaporthe grisea during infectious structure development. Mol Microbiol 73:553–570

110. Mehrabi R, Ding S, Xu J-R (2008) MADS-box transcription factor Mig1 is required for infectious growth in Magnaporthe grisea. Eukaryot Cell 7:791

111. Qi Z, Wang QI, Dou X, Wang WEI, Zhao Q, Lv R, Zhang H, Zheng X, Wang P, Zhang Z (2012) MoSwi6, an APSES family transcription factor, interacts with MoMps1 and is required for hyphal and conidial morphogenesis, appressorial function and pathogenicity of Magnaporthe oryzae. Mol Plant Pathol 13:677–689

112. Fernandez J, Wilson RA (2012) Why no feeding frenzy? Mechanisms of nutrient acquisition and utilization during infection by the rice blast fungus Magnaporthe oryzae. Mol Plant-Microbe Interact 25:1286–1293

113. Fernandez J, Marroquin-Guzman M, Wilson RA (2014) Mechanisms of nutrient acquisition and utilization during fungal infections of leaves. Annu Rev Phytopathol 52:155–174

114. Fernandez J, Marroquin-Guzman M, Wilson RA (2014) Evidence for a transketolase-mediated metabolic checkpoint governing biotrophic growth in rice cells by the blast fungus Magnaporthe oryzae. PLoS Pathog 10:e1004354

115. Sun G, Elowsky C, Li G, Wilson RA (2018) TOR-autophagy branch signaling via Imp1 dictates plant-microbe biotrophic interface longevity. PLoS Genet 14:e1007814

116. Molloy S (2010) Fungal biology: Magnaporthe effectors on the move. Nat Rev Microbiol 8:466–467

117. Kamoun S (2006) A catalogue of the effector secretome of plant pathogenic oomycetes. Annu Rev Phytopathol 44:41–60

118. Govers F, Bouwmeester K (2008) Effector trafficking: RXLR-dEER as extra gear for delivery into plant cells. Plant Cell 20:1728–1730

119. Saitoh H, Fujisawa S, Mitsuoka C, Ito A, Hirabuchi A, Ikeda K, Irieda H, Yoshino K, Yoshida K, Matsumura H, Tosa Y, Win J, Kamoun S, Takano Y, Terauchi R (2012)

Large-scale gene disruption in Magnaporthe oryzae identifies MC69, a secreted protein required for infection by monocot and dicot fungal pathogens. PLoS Pathog 8:e1002711
120. Park SY, Choi J, Lim SE, Lee GW, Park J, Kim Y, Kong S, Kim SR, Rho HS, Jeon J, Chi MH, Kim S, Khang CH, Kang S, Lee YH (2013) Global expression profiling of transcription factor genes provides new insights into pathogenicity and stress responses in the rice blast fungus. PLoS Pathog 9:e1003350
121. van der Hoorn RA, Kamoun S (2008) From guard to decoy: a new model for perception of plant pathogen effectors. Plant Cell 20:2009–2017
122. Hogenhout SA, Van der Hoorn RA, Terauchi R, Kamoun S (2009) Emerging concepts in effector biology of plant-associated organisms. Mol Plant-Microbe Interact 22:115–122
123. Birch PR, Rehmany AP, Pritchard L, Kamoun S, Beynon JL (2006) Trafficking arms: oomycete effectors enter host plant cells. Trends Microbiol 14:8–11
124. Win J, Chaparro-Garcia A, Belhaj K, Saunders DG, Yoshida K, Dong S, Schornack S, Zipfel C, Robatzek S, Hogenhout SA, Kamoun S (2012) Effector biology of plant-associated organisms: concepts and perspectives. Cold Spring Harb Symp Quant Biol 77:235–247
125. Bozkurt TO, Schornack S, Banfield MJ, Kamoun S (2012) Oomycetes, effectors, and all that jazz. Curr Opin Plant Biol 15:483–492
126. Xin X-F, Nomura K, Aung K, Velásquez AC, Yao J, Boutrot F, Chang JH, Zipfel C, He SY (2016) Bacteria establish an aqueous living space in plants crucial for virulence. Nature 539:524–529
127. Liu W, Liu J, Ning Y, Ding B, Wang X, Wang Z, Wang GL (2013) Recent progress in understanding PAMP- and effector-triggered immunity against the rice blast fungus Magnaporthe oryzae. Mol Plant 6:605–620. https://doi.org/10.1093/mp/sst015
128. Chisholm ST, Coaker G, Day B, Staskawicz BJ (2006) Host-microbe interactions: shaping the evolution of the plant immune response. Cell 124:803–814
129. Banfield MJ (2015) Perturbation of host ubiquitin systems by plant pathogen/pest effector proteins. Cell Microbiol 17:18–25
130. Oliveira-Garcia E, Valent B (2015) How eukaryotic filamentous pathogens evade plant recognition. Curr Opin Microbiol 26:92–101
131. Azizi P, Rafii MY, Abdullah SN, Nejat N, Maziah M, Hanafi MM, Latif MA, Sahebi M (2016) Toward understanding of rice innate immunity against Magnaporthe oryzae. Crit Rev Biotechnol 36:165–174
132. Bentham A, Burdett H, Anderson PA, Williams SJ, Kobe B (2016) Animal NLRs provide structural insights into plant NLR function. Ann Bot 119(5):827–702
133. Stein JC, Yu Y, Copetti D, Zwickl DJ, Zhang L, Zhang C, Chougule K, Gao D, Iwata A, Goicoechea JL, Wei S, Wang J, Liao Y, Wang M, Jacquemin J, Becker C, Kudrna D, Zhang J, Londono CEM, Song X, Lee S, Sanchez P, Zuccolo A, Ammiraju JSS, Talag J, Danowitz A, Rivera LF, Gschwend AR, Noutsos C, Wu C-c, S-m K, J-w Z, F-j W, Zhao Q, Feng Q, El Baidouri M, Carpentier M-C, Lasserre E, Cooke R, da Rosa Farias D, da Maia LC, dos Santos RS, Nyberg KG, McNally KL, Mauleon R, Alexandrov N, Schmutz J, Flowers D, Fan C, Weigel D, Jena KK, Wicker T, Chen M, Han B, Henry R, Hsing Y-iC, Kurata N, de Oliveira AC, Panaud O, Jackson SA, Machado CA, Sanderson MJ, Long M, Ware D, Wing RA (2018) Genomes of 13 domesticated and wild rice relatives highlight genetic conservation, turnover and innovation across the genus Oryza. Nat Genet 50:285–296
134. Jia Y, McAdams SA, Bryan GT, Hershey HP, Valent B (2000) Direct interaction of resistance gene and avirulence gene products confers rice blast resistance. EMBO J 19:4004–4014
135. Kanzaki H, Yoshida K, Saitoh H, Fujisaki K, Hirabuchi A, Alaux L, Fournier E, Tharreau D, Terauchi R (2012) Arms race co-evolution of Magnaporthe oryzae AVR-Pik and rice Pik genes driven by their physical interactions. Plant J 72:894–907
136. Yoshida K, Saitoh H, Fujisawa S, Kanzaki H, Matsumura H, Yoshida K, Tosa Y, Chuma I, Takano Y, Win J, Kamoun S, Terauchi R (2009) Association genetics reveals three novel avirulence genes from the Rice blast fungal pathogen Magnaporthe oryzae. Plant Cell 21:1573
137. Chuma I, Isobe C, Hotta Y, Ibaragi K, Futamata N, Kusaba M, Yoshida K, Terauchi R, Fujita Y, Nakayashiki H, Valent B, Tosa Y (2011) Multiple translocation of the AVR-Pita effector gene among chromosomes of the rice blast fungus

Magnaporthe oryzae and related species. PLoS Pathog 7:e1002147

138. Zhang S, Wang L, Wu W, He L, Yang X, Pan Q (2015) Function and evolution of Magnaporthe oryzae avirulence gene AvrPib responding to the rice blast resistance gene Pib. Sci Rep 5:11642

139. Inoue Y, Vy TTP, Yoshida K, Asano H, Mitsuoka C, Asuke S, Anh VL, Cumagun CJR, Chuma I, Terauchi R, Kato K, Mitchell T, Valent B, Farman M, Tosa Y (2017) Evolution of the wheat blast fungus through functional losses in a host specificity determinant. Science 357:80

140. De Wit PJ, Mehrabi R, Van den Burg HA, Stergiopoulos I (2009) Fungal effector proteins: past, present and future. Mol Plant Pathol 10:735–747

141. Wang B-H, Ebbole DJ, Wang Z-h (2017) The arms race between Magnaporthe oryzae and rice: diversity and interaction of Avr and R genes. J Integr Agric 16:2746–2760

142. Orbach MJ, Farrall L, Sweigard JA, Chumley FG, Valent B (2000) A telomeric avirulence gene determines efficacy for the rice blast resistance gene pi-ta. Plant Cell 12:2019–2032

143. Bohnert HU, Fudal I, Dioh W, Tharreau D, Notteghem JL, Lebrun MH (2004) A putative polyketide synthase/peptide synthetase from Magnaporthe grisea signals pathogen attack to resistant rice. Plant Cell 16:2499–2513

144. Wolpert TJ, Dunkle LD, Ciuffetti LM (2002) Host-selective toxins and avirulence determinants: what's in a name? Annu Rev Phytopathol 40:251–285

145. Berruyer R, Adreit H, Milazzo J, Gaillard S, Berger A, Dioh W, Lebrun MH, Tharreau D (2003) Identification and fine mapping of Pi33, the rice resistance gene corresponding to the Magnaporthe grisea avirulence gene ACE1. Theor Appl Genet 107:1139–1147

146. Wu CH, Abd-El-Haliem A, Bozkurt TO, Belhaj K, Terauchi R, Vossen JH, Kamoun S (2017) NLR network mediates immunity to diverse plant pathogens. Proc Natl Acad Sci U S A 114:8113–8118

147. Cesari S, Kanzaki H, Fujiwara T, Bernoux M, Chalvon V, Kawano Y, Shimamoto K, Dodds P, Terauchi R, Kroj T (2014) The NB-LRR proteins RGA4 and RGA5 interact functionally and physically to confer disease resistance. EMBO J 33:1941–1959

148. Cesari S, Thilliez G, Ribot C, Chalvon V, Michel C, Jauneau A, Rivas S, Alaux L, Kanzaki H, Okuyama Y, Morel JB, Fournier E, Tharreau D, Terauchi R, Kroj T (2013) The rice resistance protein pair RGA4/RGA5 recognizes the Magnaporthe oryzae effectors AVR-pia and AVR1-CO39 by direct binding. Plant Cell 25:1463–1481

149. Maqbool A, Saitoh H, Franceschetti M, Stevenson CEM, Uemura A, Kanzaki H, Kamoun S, Terauchi R, Banfield MJ (2015) Structural basis of pathogen recognition by an integrated HMA domain in a plant NLR immune receptor. elife 4:e08709

150. De la Concepcion JC, Franceschetti M, Maqbool A, Saitoh H, Terauchi R, Kamoun S, Banfield MJ (2018) Polymorphic residues in rice NLRs expand binding and response to effectors of the blast pathogen. Nat Plants 4:576–585

151. De la Concepcion JC, Franceschetti M, MacLean D, Terauchi R, Kamoun S, Banfield MJ (2019) Protein engineering expands the effector recognition profile of a rice NLR immune receptor. elife 8:e47713

# Chapter 3

# Analysis of the Secondary Metabolism in *Magnaporthe oryzae*


## Elizabeth Skellam



## Abstract

*Magnaporthe oryzae* produces a number of secondary metabolites, some of which are thought to be responsible for the virulence of this fungus toward rice. Due to the importance of understanding plant-pathogen interactions, several of these metabolites have been investigated chemically and biosynthetically. This chapter provides an overview of the secondary metabolites isolated from *M. oryzae* and describes a general method for metabolite extraction, followed by an analysis using high-performance liquid chromatography (HPLC) combined with mass spectrometry (LCMS).



**Key words** Secondary metabolism, Secondary metabolites, Natural products, Liquid chromatography—mass spectrometry (LCMS), Sample preparation, Liquid-liquid extraction, *Magnaporthe oryzae*, Fungi


## 1 Introduction

Secondary metabolism is the area of metabolism dedicated to the production of small molecules (natural products) that are not deemed essential for life but do offer the producing organism a competitive advantage [1]. Unlike primary metabolism that is shared across all organisms, secondary metabolism differs significantly between different species [2]. Furthermore, the full range of secondary metabolites that can be produced by an organism are not produced under all conditions; they may require specific environmental signals [3]. Of the secondary metabolites that are observed, the exact roles of the small molecules are often unknown but are considered to function as toxins, signaling molecules, pigments, or antioxidants. They appear to enable the producing organism to adapt to unique and hostile environments [4]. Many of these compounds also play an essential role in plant-pathogen interactions [5].


Stefan Jacob (ed.), *Magnaporthe oryzae: Methods and Protocols*, Methods in Molecular Biology, vol. 2356,
https://doi.org/10.1007/978-1-0716-1613-0_3,

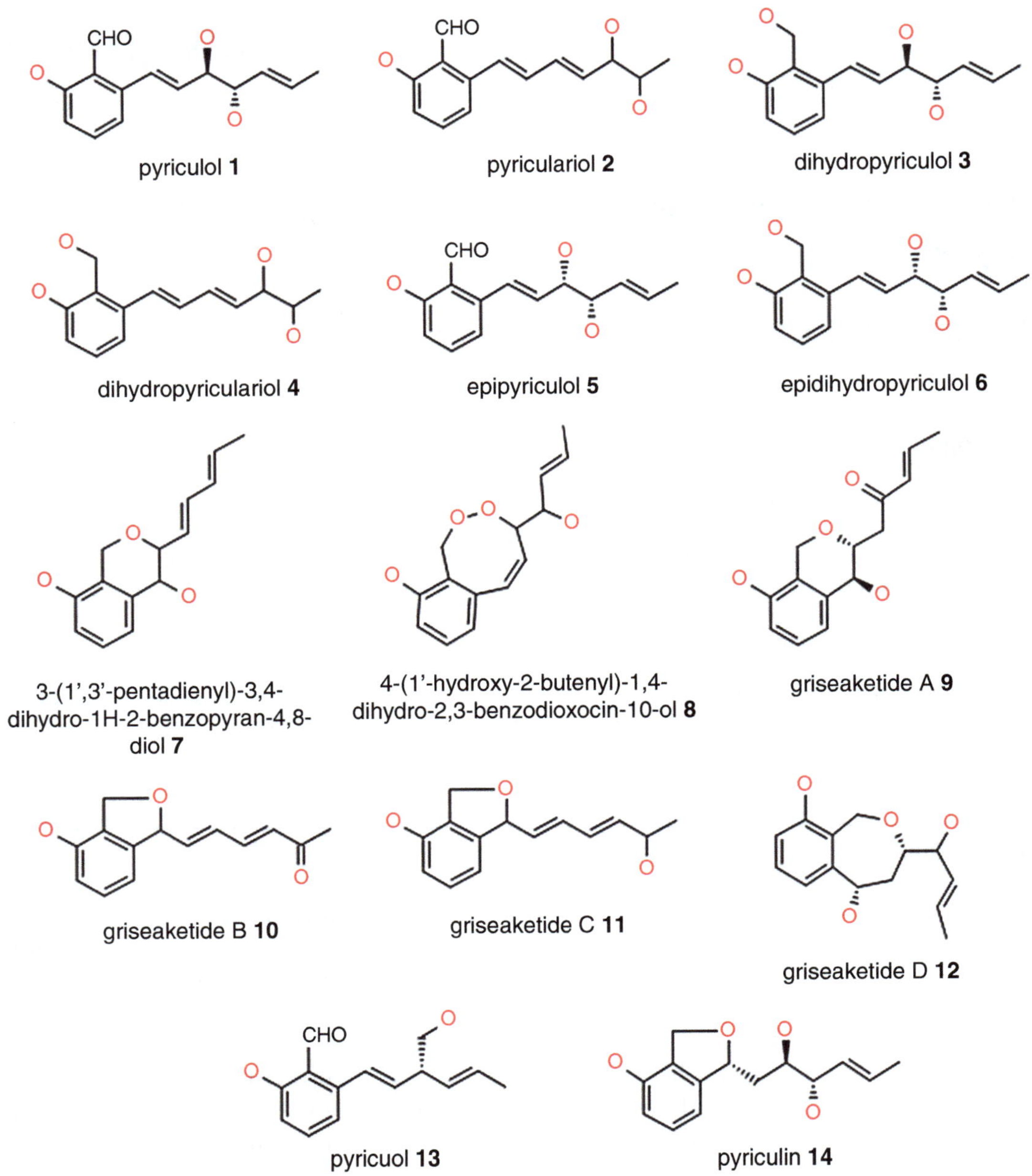


**Fig. 1** Pyriculol-related secondary metabolites isolated from *M. oryzae*

The secondary metabolism of the rice blast fungus *Magnaporthe oryzae* has been investigated over many years and revealed numerous classes of secondary metabolites. One of the earliest reports of secondary metabolites from *M. oryzae* is the polyketide pyriculol **1** [6]. Pyriculol **1** induces lesion formation on rice leaves and is thought to be involved in the virulence of *M. oryzae* against rice [6, 7]. Additional molecules **2–14** in the pyriculol family are shown in Fig. 1 [8–11].

A gene cluster encoding some of the enzymes, e.g., a polyketide synthase (MoPKS19) and an oxidoreductase (MoC19OXR1) required for pyriculol biosynthesis, was confirmed by gene inactivation studies; however, the exact sequence of reactions leading to this family of molecules still needs to be elucidated [7]. The recent characterization of the biosynthetic pathway toward closely related salicylaldehydes in *Trichoderma virens* may help identify other enzymes involved [12]. Biological testing of the *M. oryzae* mutant strains unable to produce pyriculol **1** and pyriculariol **2** was still pathogenic toward rice, indicating that these secondary metabolites are not required for infection [7].

Another family of secondary metabolites extensively studied from *M. oryzae* belong to the dihydroxynaphthalene (DHN) melanin pathway. Melanin is a multifunctional pigment providing protection from ultraviolet light, ionizing radiation, and oxidizing agents as well as having a role in pathogenesis [13]. Isolated intermediates include scytalone **15** and vermelone **16** [14–17] and characterized off-pathway shunt metabolites including flaviolin **17**, 3,4,8-trihydroxy-1(2*H*)-napthalone **18**, 2-hydroxyjuglone **19**, and isosclerone **20** (Fig. 2) [15, 18].

Additional families of polyketide-derived secondary metabolites include 6-hydroxymellein **21**, the related isocoumarin O-demethyldiaporthin **22** [19], and the phytotoxin terrestric acid **23** (Fig. 3) [20]. The biosynthetic gene clusters encoding 6-hydroxymellein and terrestric acid have been elucidated in other fungi [21–23].

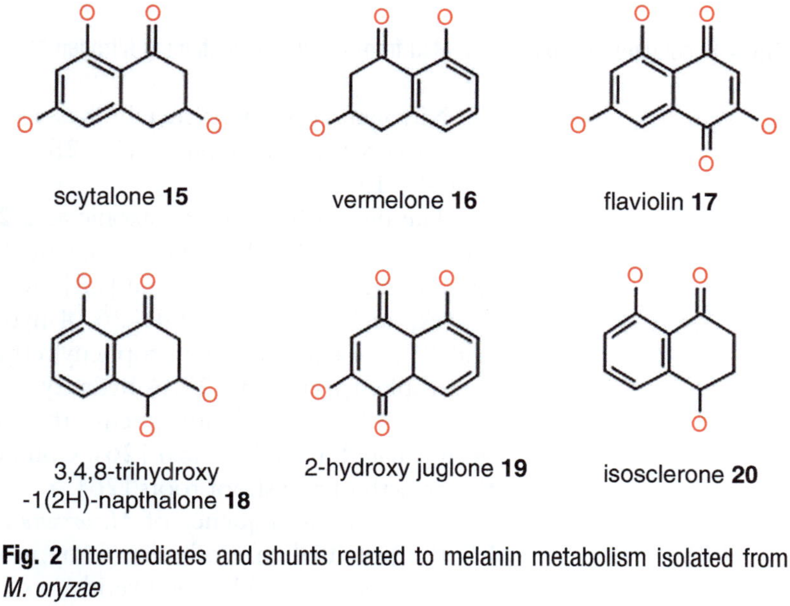


**Fig. 2** Intermediates and shunts related to melanin metabolism isolated from *M. oryzae*

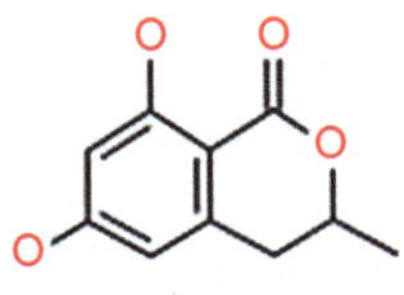

6-hydroxymellein **21**

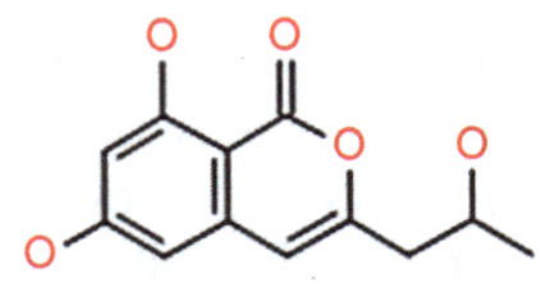

O-demethyldiaporthin **22**

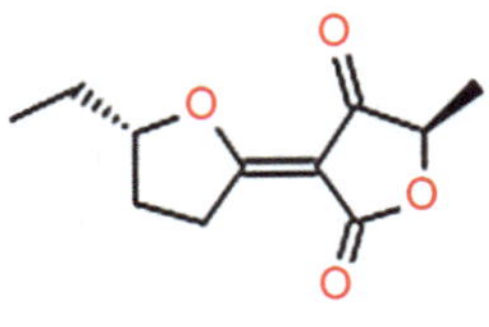

terrestric acid **23**

**Fig. 3** Miscellaneous polyketides isolated from *M. oryzae*

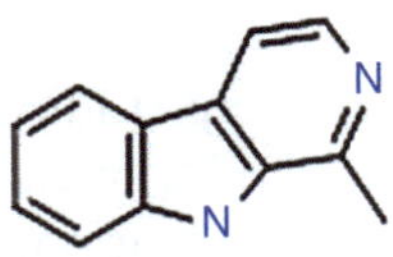

harmane **24**

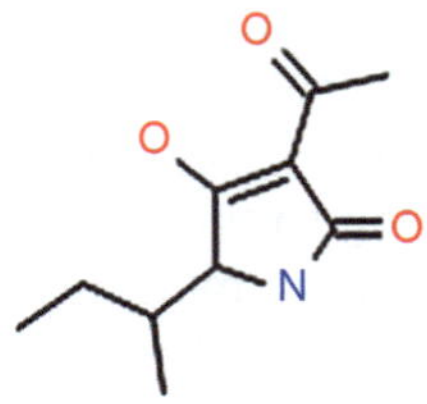

tenuazonic acid **25**

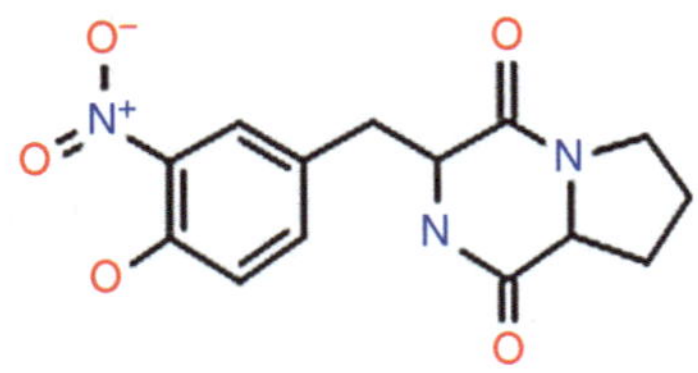

pyriculamide **26**

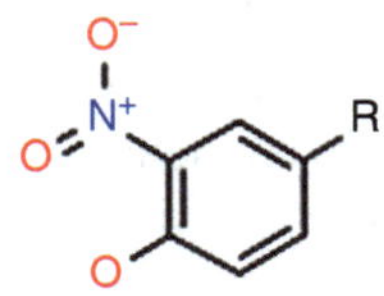

4-hydroxy-3-nitrophenyl acetic acid **27** (R = $CH_2COOH$)
2-(4-hydroxy-3-nitrophenyl) ethanol **28** (R = CH(OH)Me)
1-(4-hydroxy-3-nitrophenyl) ethanol **29** (R = $CH_2CH_2OH$)
4-hydroxy-3-nitrobenzyl alcohol **30** (R = $CH_2OH$)
N-(4-hydroxy-3-nitrophenylethyl)acetamide **31** (R = $CH_2CH_2NHAc$)

**Fig. 4** Secondary metabolites isolated from *M. oryzae* containing nitrogen atoms

Nitrogen-containing metabolites including harmane **24** and the mycotoxin tenuazonic acid **25** have also been reported (Fig. 4) [19, 24].

The biosynthesis of tenuazonic acid **25** was shown to require the unusual NRPS-PKS hybrid enzyme TAS1 that utilizes isoleucine and acetoacetyl-coenzyme [25]. Several O-nitrophenol derivatives, such as pyriculamide **26**, 4-hydroxy-3-nitrophenylacetic acid **27**, 1-(4-hydroxy-3-nitrophenyl(ethanol) **28**, 2-(4-hydroxy-3-nitrophenyl(ethanol) **29**, 4-hydroxy- 3-nitrobenzyl alcohol **30**, and *N*-(4-hydroxy- 3-nitrophenylethyl)acetamide **31**, have also been isolated from *M. oryzae* [26]. Compounds **27** and **29** inhibit the growth of rice shoot roots [26].

The genome sequence of *M. oryzae* indicates that this fungus has the potential to synthesize many more families of secondary metabolites as 22 polyketide synthase (PKS) genes, 8 non-ribosomal peptide synthetase (NRPS) genes, 10 polyketide synthase non-ribosomal peptide synthetase (PKS-NRPS) genes, 2 terpene cyclase genes, and 3 dimethylallyl tryptophan synthase (DMATS) genes have been predicted [27]. Three PKS genes, MGG_04775, MGG_07219, and MGG_10072, were shown to be upregulated during infection-related development [28]. The ACE1 gene cluster has been identified as being expressed specifically during *M. oryzae* penetration of rice leaves, and although the structures of the secondary metabolites encoded by this gene cluster remain cryptic, the function of one metabolite is known to be responsible for a signal that alerts resistant strains of rice to activate a defense response [29–31].

Several attempts have been made to identify the cryptic secondary metabolites encoded by the ACE1 gene cluster using methods that bypass the tight regulation of this cluster. The gene cluster encodes two PKS-NRPS enzymes, ACE1 and SYN2; two *trans*-acting enoylreductases (ER), RAP1 and RAP2; four cytochrome P450 monooxygenases, CYP1–4; two oxidoreductases, OXR1 and OXR2; a Diels-Alderase ORF3; a hydrolase, ORFZ; an *O*-methyltransferase, OME1; a transporter, MFS1; and a transcription factor, BC2 [31]. Heterologous gene expression of the two PKS-NRPS genes, *ACE1* and *SYN2*, with their *trans*-acting ER partners, *RAP1* and *RAP2*, led to the elucidation of a putative cytochalasan precursor **32** and a decalin molecule **33** (Fig. 5) [32–34]. The functions of the four P450 genes and the Diels-Alderase were investigated using combinatorial biosynthesis in the closely related host *M. grisea* NI980 [35, 36]. Overexpression of the transcription factor BC2 unexpectedly led to the production of hinnulin **34** postulated to be due to the unintentional activation of MGG_07219 [37].

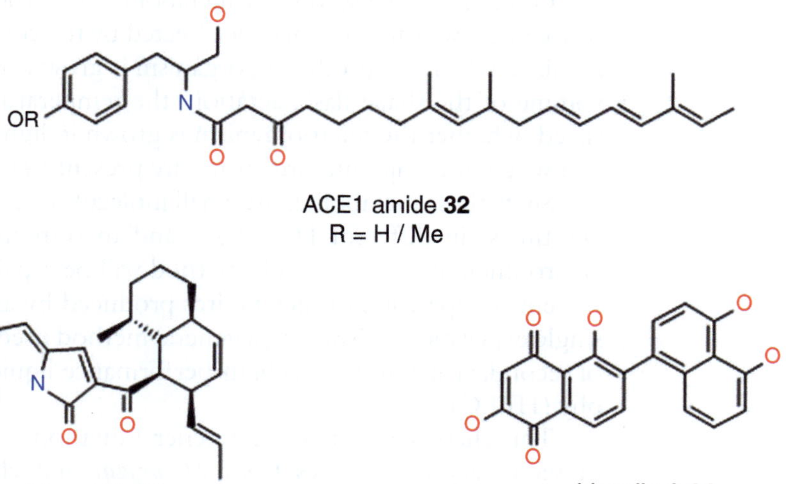


ACE1 amide **32**
R = H / Me

niduporthin **33**

hinnulin A **34**

**Fig. 5** Secondary metabolites isolated from studies of the cryptic ACE1 biosynthetic gene cluster present in *M. oryzae*

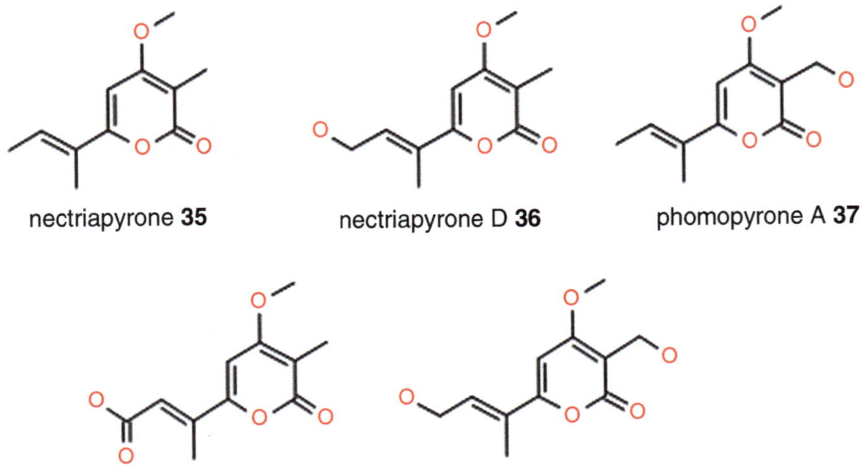

nectriapyrone **35** nectriapyrone D **36** phomopyrone A **37**

acropyrone **38** zaepyrone **39**

**Fig. 6** Pyrones isolated after activating the cryptic NEC1/2 biosynthetic genes present in *M. oryzae*

Alternative methods to induce secondary metabolite production have included disrupting the two-component signal transduction system [25, 38]. Through disruption of the high-osmolarity glycerol mitogen-activated protein (MAP) kinase gene *OSM1* and the histidine-containing phosphotransfer protein gene *PoYPD1*, the gene cluster for nectriapyrone **35** was identified [39]. Overexpression of the polyketide synthase gene *NEC1* and the *O*-methyltransferase gene *NEC2* led to the production of several nectriapyrone analogues **36–39** (Fig. 6).

Nectriapyrone mutant strains were shown to still be as virulent to rice strains compared to the wildtype, indicating that nectriapyrones are not involved in infection. Nectriapyrone **35** did however effect growth and pigment formation of *Streptomyces griseus* [39].

Often microbial secondary metabolites are produced in the lag phase of growth and may also be affected by the composition of the media in which the producing organism is grown (Fig. 7) [40], the volume of the shake flask, aeration, the temperature, the rotation speed, whether the microorganism is grown in light or in the dark, and whether competing organisms are present [41].

Since these metabolites are small molecules, produced at different times, in different life stages, and in response to unknown environmental cues, no single method will be capable of analyzing the entire repertoire of metabolites produced by an organism in a single experiment. However, a general method used for the analysis of secondary metabolites is high-performance liquid chromatography (HPLC).

This chapter will describe a general method for the extraction of secondary metabolites from *M. oryzae* and chemical analysis using liquid chromatography coupled with mass spectrometry (LCMS). LCMS is selected, because it is a very sensitive technique

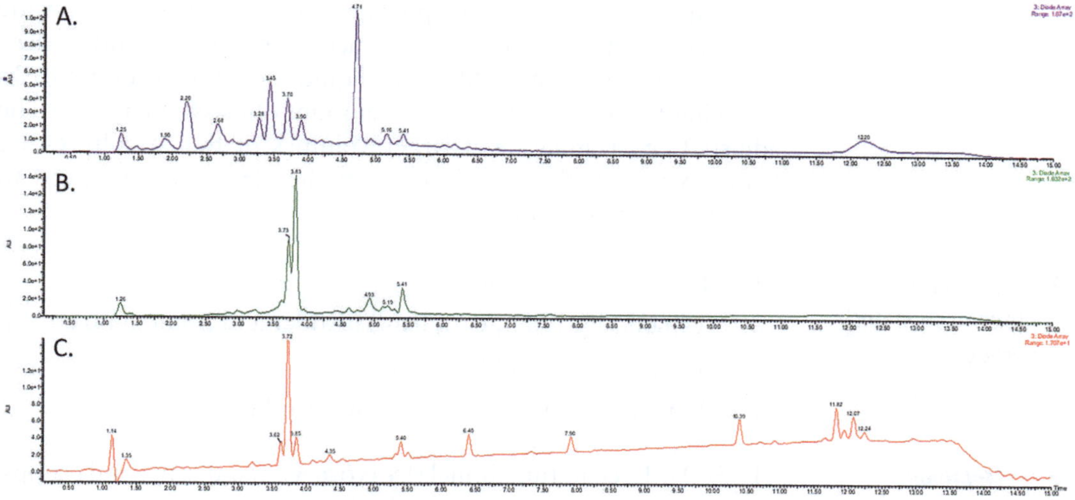


**Fig. 7** HPLC trace (detected using a diode array detector (DAD)) of *M. oryzae* Guy11 cultured using different conditions; (**a**) *M. oryzae* Guy11 grown in dextrin-peptone-yeast (DPY) extract medium for 4 days. (**b**) *M. oryzae* Guy11 grown in soy sauce sucrose (SSS) medium for 4 days. (**c**) *M. oryzae* Guy11 grown on oatmeal agar for 12 days. Each chromatographic peak represents a metabolite and each metabolite has a defined retention time, UV absorbance, and mass

that enables the separation and detection of metabolites with a broad range of chemical functionalities and molecular masses. Furthermore, LCMS can be scaled up so that the crude extract can be separated into fractions that contain enough material that full structural elucidation of individual secondary metabolites is possible (e.g., collection of mg quantities for NMR analysis). This methodology can be applied to study the secondary metabolites produced by other filamentous fungi.

## 2 Materials

There are numerous media and culture conditions that could be investigated for secondary metabolite production and so are out of scope of this protocol (*see* **Note 1**). It is important to note that more specialized extraction (e.g., hydrophobic crosslinked polystyrene copolymer XAD resins) and separation procedures may be required for enrichment of low-abundant secondary metabolites, or metabolites that are especially polar/non-polar/charged. Utilizing several detectors simultaneously increases the likelihood that the metabolite of interest is detected and ensures that maximum information about chemical properties is gathered.

Generally, secondary metabolite extraction is simpler if the fungi are grown in aerated shake flasks; however, this is not always the most suitable method. Furthermore, if the ultimate aim of the experiment is to isolate and structurally characterize secondary metabolites, liquid culture volumes may need to be scaled up. This protocol describes preparation and analysis of a crude secondary metabolite mixture extract from one 500 mL Erlenmeyer flask containing 100 mL of liquid medium. If several flasks are being extracted in parallel, the amount of glassware needs to be adjusted accordingly, or if larger volumes of fungal cultures are being extracted, the specifications of the glassware used need to be considered (*see* **Note 2**).

### 2.1 Mycelium Homogenization and Separation

1. Glass beaker (*see* **Note 3**).
2. Mechanical grinder, e.g., hand blender and homogenizer.
3. Buchner funnel and filter paper.
4. 500 mL vacuum flask, tubing, seal, and pump.

### 2.2 Metabolite Extraction

1. 500 mL separating funnel (*see* **Note 4**), stand, ring, and clamps.
2. Solvents, e.g., ethyl acetate.
3. 1 M hydrochloric acid/1 M sodium hydroxide, pH indicator paper, glass pipettes, and rubber bulb.
4. Graduate glass cylinder (>100 mL).
5. 500 mL Erlenmeyer flask.
6. 2× glass filter funnel.
7. Filter paper.
8. $MgSO_4$ and spatula.
9. 500 mL round-bottomed flask (*see* **Note 5**).
10. Analytical balance (to four decimal places).

### 2.3 Sample Preparation

1. Rotary evaporator and vacuum pump.
2. Solvents, e.g., methanol.
3. Glass pipettes, rubber bulb, and glass wool.
4. Glass vials, lids, and sticky paper labels.
5. Glass vials with septum lids suitable for automated injection into LCMS (*see* **Note 6**).

### 2.4 High-Performance Liquid Chromatography

This protocol describes the use of a Waters mass-directed autopurification system comprising of a Waters 2767 autosampler, a Waters 2545 pump system, a Waters 2424 evaporative light scattering detector (ELSD), a Waters 2998 diode array detector (DAD), and a Waters SQD-2 mass spectrometer (MS) that utilizes electrospray ionization (ESI) (*see* **Note** 7). Control of the system is via the software program MassLynx.

1. Mobile phase A: double distilled deionized water + 0.05% formic acid (*see* **Note 8**).
2. Mobile phase B: HPLC grade acetonitrile + 0.045% formic acid.

3. C18 reversed-phase silica stationary phase (dimensions 4.6 × 100 mm; particle size 2.6 μm) for analytical HPLC, e.g., Phenomenex Kinetex column (*see* **Note 9**).
4. C18 reversed-phase silica stationary phase (dimensions 21.2 × 250 mm; particle size 5 μm for semi-preparative scale HPLC, e.g., Phenomenex Kinetic Axia (*see* **Note 9**)).
5. Glass test tubes.
6. Vacuum centrifuge.

## 3 Methods

The general workflow for extracting metabolites from fungi and separating into individual fractions is shown in Fig. 8.

Perform all extractions and handling of chemicals in a chemical fume hood using clean glassware; do not use plastics. Organic solvents and potential mycotoxins are a health hazard, so handle them in a fume hood with appropriate personal protective equipment. Consult safety data sheets before handling chemicals and attend to spillages immediately. Dispose solvents and waste according to local authority guidelines. Inactivate fungal material that has not been in contact with solvents by autoclaving in liquid cycle (121 °C, 15 psi) for at least 30 min (*see* **Note 10**). Clean surfaces that have been in contact with fungal material, using a disinfectant.

### 3.1 Mycelium Homogenization and Separation

Not all secondary metabolites are excreted into the supernatant, meaning that the secondary metabolite profile can differ between the mycelial extract and the supernatant extract. To ensure that as many metabolites can be extracted in as high concentrations as possible, blend the mycelium to release the metabolites into the supernatant (*see* **Note 11**).

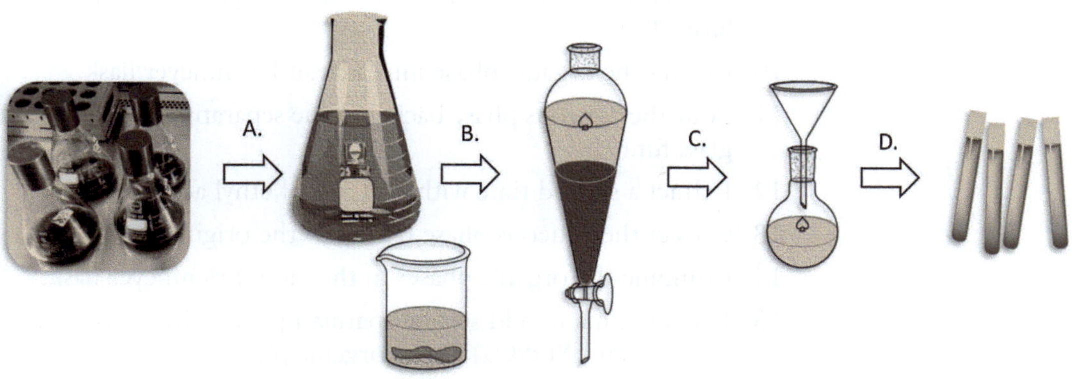


**Fig. 8** General workflow for extracting secondary metabolites from fungi and fractionating via preparative HPLC. (**a**) Mycelium homogenization or separation. (**b**) Liquid–liquid extraction of supernatant. (**c**) Concentration of extract. (**d**) Resulting fractions collected after separation by HPLC

1. Pour contents of fermentation flask(s) into the glass beaker.
2. Use the hand blender/homogenizer following the manufacturer's instructions to break up mycelial clumps and generate a smooth uniform sample.
3. Assemble vacuum filtration equipment.
4. Place the filter paper disc inside the Buchner funnel and moisten with a few drops of water to seal the filter paper to the funnel.
5. Pour the homogenized fungal culture into the Buchner funnel to separate the disrupted mycelium from the supernatant.
6. When all of the filtrate is collected in the filter flask, remove the filter paper and the cells from the funnel and dispose of appropriately.

### 3.2 Metabolite Extraction

1. Secure the separating funnel and ensure that the tap is closed.
2. Place a clean empty Erlenmeyer flask underneath the separating funnel.
3. Adjust the pH of the supernatant as appropriate (*see* **Note 12**).
4. Pour the supernatant into the separating funnel via a clean glass funnel. Pouring via a funnel prevents potential mycotoxins dripping on the outside of the glassware.
5. Measure approx. 100 mL of ethyl acetate using a graduated cylinder and add to the separating funnel via the same glass funnel.
6. Add a stopper to the separating funnel and invert several times. Vent if necessary.
7. Swirl the two phases vigorously to promote mixing while avoiding the potential for an emulsion.
8. Let the two layers separate.
9. Collect the aqueous phase into the original filter flask (*see* **Note 13**).
10. Collect the organic phase into a clean Erlenmeyer flask.
11. Pour the aqueous phase back into the separating funnel via the glass funnel.
12. Extract a second time with 100 mL of ethyl acetate.
13. Collect the aqueous phase back into the original filter flask.
14. Combine the organic phases in the same Erlenmeyer flask.
15. Use a spatula to add several spatula tips of anhydrous magnesium sulfate ($MgSO_4$) to the organic phase.
16. Remove the $MgSO_4$ by gravity filtration into a clean, dry, pre-weighed 250 mL round-bottomed flask (*see* **Note 5**).
17. Remove the solvent by rotary evaporation, and when you have a dry fungal extract, let the flask cool to room temperature.

### 3.3 Sample Preparation

1. Weigh the crude fungal extract and add methanol (MeOH) to make the crude extract to 10 mg/mL. Do not inject higher concentrations into the LCMS as detectors (e.g., DAD, MS) become overloaded.
2. Add a small amount of cotton/glass wool to the inside of a glass Pasteur pipette. A second glass Pasteur pipette can be used to ensure the wool plug reaches the constriction.
3. Filter the crude extract through the plugged Pasteur pipette and collect in an appropriate glass vial (*see* **Note 14**). The extract should be clear with no debris as the particles can cause the injection needle to clog.
4. If necessary transfer to a HPLC vial. If it is not analyzed by HPLC, store the extract in a cool dark place or freeze at −20 °C (*see* **Note 15**).

### 3.4 High-Performance Liquid Chromatography

1. Check that solvent reservoirs contain enough solvents and that the waste solvent chamber is not too full.
2. Inject blank solvents (e.g., MeCN, $H_2O$) and observe whether the system pressure fluctuates. No peaks should be observed in the resultant chromatogram. The system pressure should not be greater than recommended by the manufacturer.
3. Monitor absorbance from 200 to 600 nm using the Waters 2998 diode array detector (DAD).
4. Use the following conditions for a Waters SQD2 MS: desolvation temperature 350 °C, desolvation gas flow rate 800 L/h, cone gas flow rate 50 L/h, and source temperature 150 °C (*see* **Note 16**).
5. HPLC gradients and flow rates used are described in Table 1.
6. Perform an analytical run by injecting up to 20 μL of crude extract via the autosampler into the LCMS column. Molecules with a chromophore will be detected by the DAD; nonvolatile molecules will be detected by the ELSD; and ionizable molecules will be detected by the MS (*see* **Note 17**).

**Table 1**
**Standard gradient elution timetable for mobile phases A (water + 0.05% formic acid) and B (acetonitrile +0.045% formic acid). Flow rate = 1 mL/min (analytical) or 20 mL/min (prep)**

| Time (min) | % A | % B |
|---|---|---|
| 0 | 90 | 10 |
| 10 | 10 | 90 |
| 12 | 10 | 90 |
| 15 | 90 | 10 |

7. Fractionation and purification of metabolites can be achieved using the Phenomenex Luna 10 × 250 cm 5 μm (ODS) reversed-phase semi-preparative column with the same Waters mass-directed autopurification system. The gradient and flow rate are described in Table 1 and the parameters for the DAD, ELSD, and MS remain unchanged. Fractions separated can be collected based on their mass, UV absorption, or retention time as required. For example, a fraction putatively containing tenuazonic acid (based on MS data and UV absorption data) elutes at 5.4 min (Fig. 9). Via MassLynx the following parameters can be inputted for the mass-directed purification of the metabolite of interest: Primary trigger = Mass A; ES- ion adduct = −1; ES+ ion adduct = 1, 23 ($Na^+$-adduct); adjust the minimum/maximum intensity threshold and fraction width accordingly. Mass $A$ = 197.5. Repeated injections of the crude extract (200–400 μL/injection) enable the collection of this specific metabolite of interest.
8. The collected fractions are pooled and freeze dried.
9. The purity of the sample can be confirmed by reinjection into the LCMS (Fig. 9).
10. The identity of the molecule can be confirmed by high-resolution mass spectrometry (LCMS), 1H NMR, 13C NMR, and 2D NMR experiments if necessary.

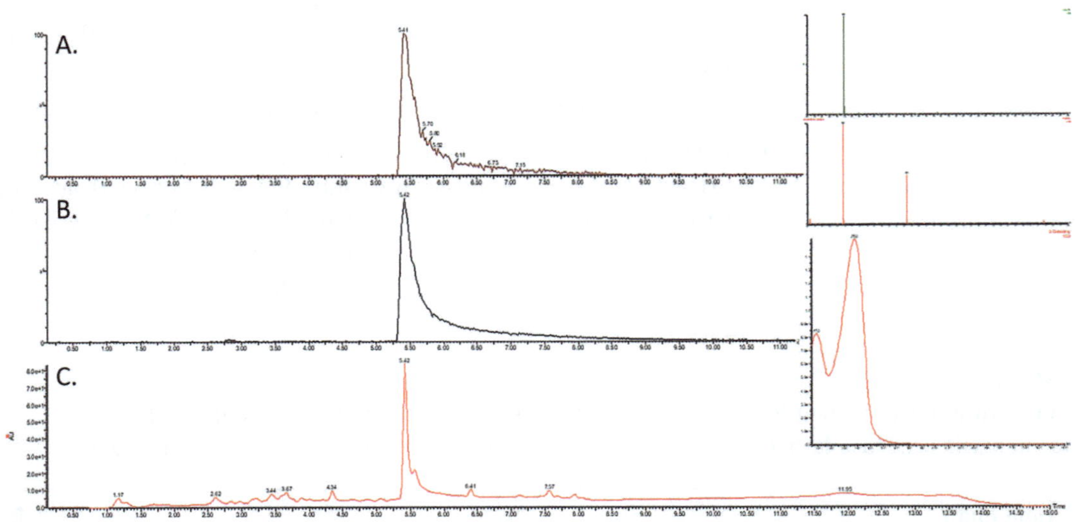


**Fig. 9** LCMS trace of purified fraction collected at 5.4 min. (**a**) Extracted ion chromatogram ($m/z$ = 196) of the ES-MS. Extracted ion chromatogram ($m/z$ = 198) of the ES+ MS. (**c**) DAD trace of reinjected fraction. Mass spectrum of compound at 5.4 min (top right) indicates molecular ion peak = 197 g/mol and UV absorbance of compound at 5.4 min (bottom right) identifies an absorbance maximum at 276 nm

## 4 Notes

1. For a typical workflow, inoculate spores onto an agar plate (e.g., oatmeal agar [4% (w/v) oatmeal, 0.5% (w/v) sucrose, 2% (w/v) agar, Millipore water]) and incubate at 25 °C for 7–10 days. From this plate, make a mycelia/spore suspension which has adequate fungal material to inoculate 5–10,500 mL Erlenmeyer flasks. Store excess spore suspension in a 50% glycerol solution at -80 °C. The Erlenmeyer flasks contain 100 mL liquid medium (e.g., soy sauce sucrose media [5% (v/v) soya sauce, 5% (w/v) sucrose, tap water]). Shake them at 110 rpm for 7–14 days at 25 °C. Alternatively, use solid media [e.g., rice, 40% (w/v) full-grain rice, soaked overnight prior to being autoclaved] and incubate the cultures without shaking at 25 °C.
2. A common procedure when screening media or investigating the optimal time to extract is to inoculate several flasks from the same agar plate (*see* **Note 1**). As each flask is being investigated separately, then it needs to be extracted separately. If only one condition is being investigated but at on a scale larger than 100 mL (e.g., 1 L or 10 × 100 mL combined), then larger glassware vessels will be required.
3. The volume of the beaker depends on the volume of the culture being extracted and the diameter of mechanical disrupter. A 500 mL glass beaker can usually accommodate a handheld blender.
4. The separating funnel should not be more than two thirds full. If extracting larger volumes, a larger separating funnel is required or the supernatant will have to be extracted in batches.
5. The weight of the empty round-bottomed flask can be written in pencil inside the neck of the flask. When filling the round-bottomed flask, the volume of liquid added should not be more than half the total volume of the flask. If extracting larger volumes, a larger round-bottomed flask is required or the extract will have to be concentrated in batches.
6. Depending on the LCMS instrument and the volume of crude extract (e.g., <100 μL), use glass inserts so that the injection needle is able to collect adequate sample for injection into the HPLC column.
7. When using other systems, ensure that the ionization source is compatible with the metabolite of interest.
8. When adding formic acid to solvents, ensure that the solutions are mixed well. Avoid using plastic pipette tips for measuring or transferring solvents.

9. The stationary phase must be able to retain the metabolites of interest. To preserve columns and protect against blockages, use guard columns or frit filters. Check the pre-columns and frits, and clean and replace them regularly.
10. Never autoclave fungal material or aqueous phases, which have come into contact with flammable solvents. Instead, treat and dispose them as chemical waste according to institutional guidelines.
11. Non-polar metabolites, e.g., fatty acids, may accumulate in the cells, and so it may be pertinent to extract the cells separately from the supernatant to avoid diluting the cell-associated metabolites.
12. The pH of the supernatant may need to be adjusted to increase extraction efficiency depending on which molecules are of interest. A lower pH is necessary when extracting acidic molecules and a higher pH is required when extracting basic molecules.
13. The solvent phase and aqueous phase separate according to density. When using ethyl acetate, this is less dense than water so it will be the top layer. If you substitute different solvents for extraction, the aqueous phase may not always be the bottom layer.
14. The HPLC vials used in this protocol hold up to 2 mL of crude extract. If more than this will be required for further analysis/purification, the extract can be kept in larger glass vials (e.g., 5 mL, 8 mL, 20 mL, etc.). It is best practice to label tubes with paper sticky labels with the identifier written in pencil. Using pen or writing on the vial directly should be avoided as if the sample is split during transfer, this can obscure the sample identifier. If the sample is presumed to be light-sensitive, the vial can be wrapped in aluminum foil and stored in a cool dark place.
15. Ideally, analyze the samples by LCMS immediately rather than store. Storage may lead to metabolites degrading over time or interacting with other molecules in the sample.
16. If using other LCMS systems, optimize the electrospray source settings following the manufacturers' instructions. The ionization source should be clean and clear of contaminants and the flow rate must be compatible with the ionization source. The ESI source must be vented out of lab.
17. Molecules may ionize in positive or negative ionization modes or both. Formation of adducts is possible and useful for analysis, e.g., addition of $Na^+$, $K^+$ in positive ion mode, addition of $HCO_2^-$ in negative ion mode, neutral losses, etc. Due to ESI being a soft ionization technique, fragmentation is not severe and often the molecular ion is straightforward to identify.

## References


1. Brakhage AA (2013) Regulation of fungal secondary metabolism. Nat Rev Microbiol 11:21–32
2. Frisvad JC, Andersen B, Thrane U (2008) The use of secondary metabolite profiling in chemotaxonomy of filamentous fungi. Mycol Res 112:231–240
3. Lim FY, Keller NP (2014) Spatial and temporal control of fungal natural product synthesis. Nat Prod Rep 31:1277–1286
4. Keller NP (2019) Fungal secondary metabolism: regulation, function and drug discovery. Nat Rev Microbiol 17:167–180
5. Zeilinger S, Gupta VK, Dahms TES, Silva RN, Singh HB, Upadhyay RS, Gomes EV, Tsui CKM, Nayak SC (2016) Friends or foes? Emerging insights from fungal interactions with plants. FEMS Microbiol Rev 40:182–207
6. Sato Z, Kozaka T (1969) Isolation and structural elucidation of a phytotoxic substance produced by *Pyricularia oryzae* cavara. Tet Lett 45:3977–3980
7. Jacob S, Grötsch T, Foster AJ, Schüffler A, Rieger PH, Sandjo LP, Liermann JC, Opatz T, Thines E (2017) Unravelling the biosynthesis of pyriculol in the rice blast fungus *Magnaporthe oryzae*. Microbiology 163:541–553
8. Kono Y, Sekido S, Yamaguchi I, Kondo H, Suzuki Y, Neto GC, Sakurai A, Yaegashi H (1991) Structures of two novel Pyriculol-related compounds and identification of naturally produced Epipyriculol from *Pyricularia oryzae*. Agric Biol Chem 55:2785–2791
9. Nukina M (1999) The blast disease fungi and their metabolic products. J Pest Sci 24:293–298
10. Kim JC, Min JY, Kim HAT, Cho KY, Yu SH (1998) Pyricuol, a new Phytotoxin from *Magnaporthe grisea*. Biosci Biotech Biochem 62:173–174
11. Yang YH, Yang DS, Lei HM, Li CY, Li GH, Zhao PJ (2020) Griseaketides A–D New Aromatic Polyketides from the Pathogenic Fungus *Magnaporthe grisea*. Molecules 25:72
12. Liu L, Tang MC, Tang Y (2019) Fungal highly reducing polyketide synthases biosynthesize salicylaldehydes that are precursors to epoxycyclohexenol natural products. J Am Chem Soc 141:19538–19541
13. Eisenman HC, Casadevall A (2011) Synthesis and assembly of fungal melanin. Appl Microbiol Biotechnol 93:931–940
14. Yamaguchi I, Sekido S, Misato T (1982) The effect of non-fungicidal anti-blast chemicals on the melanin biosynthesis and infection by *Pyricularia oryzae*. J Pesticide Sci 7:523
15. Yamaguchi I, Sekido S, Misato T (1983) Inhibition of Appressorial Melanization in *Pyricularia oryzae* by non-fungicidal anti-blast chemicals. J. Pesticide Sci 8:229
16. Thompson JE, Fahnestock S, Farrall L, Liao DI, Valent B, Jordan DB (2000) The second Naphthol reductase of fungal melanin biosynthesis in *Magnaporthe grisea*. J Biol Chem 275:34867–34872
17. Kurahashi Y, Araki Y, Kinbara T, Pontzen R, Yamaguchi I (1998) Intermediates accumulation and inhibition sites of Carpropamid in the melanin biosynthesis pathway of *Pyricularia oryzae*. J. Pesticide Sci 23:22–28
18. Maeda K, Izawa M, Nakajima Y, Jin Q, Hirose T, Nakamura T, Koshino H, Kanamara K, Ohsato S, Kamakura T, Kobayashi T, Yoshida M, Kimura M (2017) Increased metabolite production by deletion of an HDA1-type histone deacetylase in the phytopathogenic fungi, *Magnaporthe oryzae* (*Pyricularia oryzae*) and *Fusarium asiaticum*. Lett Appl Microbiol 65:446–452
19. Sviridov SI (1991) Secondary metabolites *Pyricularia oryzae* II. Polyketide metabolites. Chem Nat Compd 27:410–413
20. Nukina M (1988) Terrestric acid as a phytotoxic metabolite from *Pyricularia oryzae* cavara. Agric Biol Chem 52:2357–2358
21. Zaehle C, Gressler M, Shelest E, Geib E, Hertweck C, Brock M (2014) Terrein biosynthesis in *Aspergillus terreus* and its impact on phytotoxicity. Chem Biol 21:719–731
22. Bai T, Quan Z, Zhai R, Awakawa T, Matsuda Y, Abe I (2018) Elucidation and heterologous reconstitution of Chrodrimanin B biosynthesis. Org Lett 20:7504–7508
23. Fan J, Liao G, Ludwig-Radke L, Yin WB, Li SM (2020) Formation of Terrestric acid in *Penicillium crustosum* requires redox-assisted decarboxylation and stereoisomerization. Org Lett 22:88–92
24. Umetsu N, Kaji J, Tamari K (1972) Investigation on the toxin production by several blast fungus strains and isolation of Tenuazonic acid as a novel toxin. Agricultural Biol Chem 36:859–866
25. Yun CS, Motoyama T, Osada H (2015) Biosynthesis of the mycotoxin tenuazonic acid by a fungal NRPS–PKS hybrid enzyme. Nat Commun 6:8758

26. Sviridov SI, Ermolinkskii (1990) Secondary metabolites *Pyricularia oryzae*. Chem Nat Compd 26:691–696
27. Collemare J, Billard A, Boehnert HU, Lebrun MH (2008) Biosynthesis of secondary metabolites in the rice blast fungus *Magnaporthe grisea*: the role of hybrid PKS-NRPS in pathogenicity. Mycol Res 112:207–215
28. Dean RA, Talbot NJ, Ebbole DJ, Farman ML, Mitchell TK, Orbach MJ, Thon M, Kulkarni R, Xu JR, Pan H, Read ND, Lee YH, Carbone I, Brown D, Oh YY, Donofrio N, Jeong JS, Soanes DM, Djonovic S, Kolomiets E, Rehmeyer C, Li W, Harding M, Kim S, Lebrun MH, Bohnert H, Coughlan S, Butler J, Calvo S, Ma LJ, Nicol R, Purcell S, Nusbaum C, Galagan JE, Birren BW (2005) The genome sequence of the rice blast fungus *Magnaporthe grisea*. Nature 434:980–986
29. Böhnert HU, Fudal I, Dioh W, Tharreau D, Notteghem JL, Lebrun MH (2004) A putative polyketide synthase/peptide synthetase from *Magnaporthe grisea* signals pathogen attack to resistant rice. Plant Cell 16:2499–2513
30. Fudal I, Collemare J, Böhnert HU, Melayah D, Lebrun MH (2007) Expression of *Magnaporthe grisea* avirulence gene ACE1 is connected to the initiation of appressorium-mediated penetration. Eukaryot Cell 6:546–554
31. Collemare J, Pianferri M, Houlle AE, Morin D, Camborde L, Gaey MJ, Barbisan C, Fudal I, Lebrun MH, Böhnert HU (2008) *Magnaporthe grisea* avirulence gene ACE1 belongs to an infection-specific gene cluster involved in secondary metabolism. New Phytol 179:196–208
32. Song Z, Bakeer W, Marshall JW, Yakasai AA, Khalid RM, Collemare J, Skellam E, Tharreua D, Lebrun MH, Lazarus CM, Baily AM, Simpson TJ, Cox RJ (2015) Heterologous expression of the avirulence gene ACE1 from the fungal rice pathogen *Magnaporthe oryzae*. Chem Sci 6:4837–4845
33. Wang C, Hantke V, Cox RJ, Skellam E (2019) Targeted gene inactivations expose silent cytochalasans in *Magnaporthe grisea* NI980. Org Lett 21:4163–4167
34. Nielsen ML, Isbrandt T, Petersen LM, Mortensen UH, Andersen MR, Hoof JB, Larsen TO (2016) Linker flexibility facilitates module exchange in fungal hybrid PKS-NRPS engineering. PLoS One 11:e0161199
35. Wang C, Becker K, Pfütze S, Kuhnert E, Stadler M, Cox RJ, Skellam E (2019) Investigating the function of cryptic cytochalasan cytochrome P450 monooxygenases using combinatorial biosynthesis. Org Lett 21:8756–8760
36. Hantke V, Skellam EJ, Cox RJ (2020) Evidence for enzyme catalysed intramolecular [4+2] Diels–Alder cyclization during the biosynthesis of pyrichalasin H. Chem Commun 56 (19):2925–2928
37. Hantke V, Wang C, Skellam EJ, Cox RJ (2019) Function of pathway specific regulators in the ACE1 and pyrichalasin H biosynthetic gene clusters. RSC Adv 9:35797–35802
38. Yun CS, Motoyama T, Osada H (2017) Regulatory mechanism of mycotoxin Tenuazonic acid production in *Pyricularia oryzae*. ACS Chem Biol 12:2270–2274
39. Motoyama T, Nogawa T, Hayashi T, Hirota H, Osada H (2019) Induction of Nectriapyrone biosynthesis in the rice blast fungus *Pyricularia oryzae* by disturbance of the two-component signal transduction system. Chembiochem 20:693–700
40. Bode HB, Bethe B, Hoefs R, Zeeck A (2002) Big effects from small changes: possible ways to explore nature's chemical diversity. Chembiochem 3:619–627
41. Marmann A, Aly AH, Lin W, Wang B, Proksch P (2014) Co-cultivation - a powerful emerging tool for enhancing the chemical diversity of microorganisms. Mar Drugs 12:1043–4065

# Chapter 4

# The Determination of Carbohydrates by High-Performance Anion-Exchange Chromatography Coupled with Pulsed Amperometric Detection (HPAEC-PAD)

**Christiane Grünewald, Stefan Bohnert, and Stefan Jacob**

## Abstract

Chromatography techniques are widely used to separate, identify, and quantify molecules depending on their physicochemical properties. Standard methods range from simple size exclusion to separation based on affinity or ion exchange. Here, we present a method for the direct analysis of carbohydrates in *Magnaporthe oryzae* using high-performance anion-exchange chromatography (HPAEC) coupled with pulsed amperometric detection (PAD). The combination of HPAEC with PAD provides the highest selectivity and sensitivity with minimal sample preparation and cleanup time. Utilizing our HPAEC-PAD approach, we obtain reliable and highly reproducible determination of carbohydrates produced as osmotic stress response by *M. oryzae*. Thus, the method described provides a fast, precise, and comprehensive analysis of stress-dependent metabolic adjustments of carbohydrates not only relevant for *M. oryzae* but also applicable in other systems.

**Key words** *Magnaporthe oryzae*, Compatible solutes, High-performance anion-exchange chromatography, Pulsed amperometric detection, HPAEC, PAD, Carbohydrate analysis, Osmotic stress

## 1 Introduction

Carbohydrates are life's most essential bioactive and information-carrying molecules [1]. These important molecules have been shown to play vital roles in the biological processes of all organisms. Carbohydrate analysis has become an ever more important challenge in medical biology in the recent years, ranging from glucose monitoring and disease diagnosis (e.g., cancer and microbial infections) to their use as drugs and for basic research on molecular mechanisms in cellular physiology [1]. Microorganisms have to cope with extracellular stress during several stages of their life cycle. In order to compensate for the extracellular osmotic potential caused by drought stress or high salt concentrations, they produce and accumulate various low-molecular substances, such as carbohydrates, which act as osmolytes within the cell [2]. The

Stefan Jacob (ed.), *Magnaporthe oryzae: Methods and Protocols*, Methods in Molecular Biology, vol. 2356, https://doi.org/10.1007/978-1-0716-1613-0_4, 

filamentous phytopathogenic fungus *Magnaporthe oryzae* produces mainly arabitol, glycerol, mannitol, and trehalose, among many more, in very low concentrations [3]. A preferably sensitive carbohydrate analysis is essential for the investigation and a deeper understanding of cellular metabolic adjustments in response to osmotic stress and fungicide stress or in different developmental stages of *Magnaporthe oryzae*. The fundamental question particularly of how molecular mechanisms of eukaryotic signaling pathways evolve to adapt toward changing environmental conditions is one of the fields recently studied urgently needing the analysis of osmolytes [4].

Carbohydrates are typically analyzed using liquid chromatography methods relying upon silica-based, polymer-based, or cation-exchange columns regarding refractive index detection. Gas-liquid or gas chromatography methods with time-consuming derivatization procedures are also used [3]. These analytical methods require attention to sample solubility, sample concentration, and derivatization procedures. All of these techniques are sensitive to eluent and sample matrix components. This usually precludes the use of gradient elution and often requires stringent sample cleanup prior to injection.

In this protocol, we present an improved chromatographic technique to separate carbohydrates, namely, high-performance anion-exchange chromatography (HPAEC) coupled with pulsed amperometric detection (PAD). This method leads to the direct quantification of carbohydrates without the need of derivatization at extremely low ("picomol") concentrations with minimal sample preparation and cleanup [5–8]. Regarding the pKa values, HPAEC takes advantage of the weak acidic nature of carbohydrates. The molecules are at least partially ionized at high pH values and, thus, can be separated by anion-exchange. The separation of reduced carbohydrates, for example, sugar alcohols, by liquid chromatography with common anion-exchange columns is very difficult. That is because they are weaker acids compared to their non-reduced counterparts. The key to success is the use of a Dionex CarboPac MA1 column to separate such extremely weak acids. This column is packed with a macroporous polymer resin providing a high ion-exchange capacity. The resin is composed of a polystyrene/divinylbenzene polymeric core and the surface is crafted with quaternary ammonium anion-exchange functional groups. As a result, weak anions bind more strongly to the column, requiring high sodium hydroxide concentrations in the elution process. Consequently, the ionization of the sugar alcohols increases, which leads to good retention and high resolution on the column. Non-reduced carbohydrates can also be analyzed [9].

The PAD is a powerful detection method for carbohydrates that requires no sample derivatization. A certain potential is applied between a gold working electrode and a reference electrode

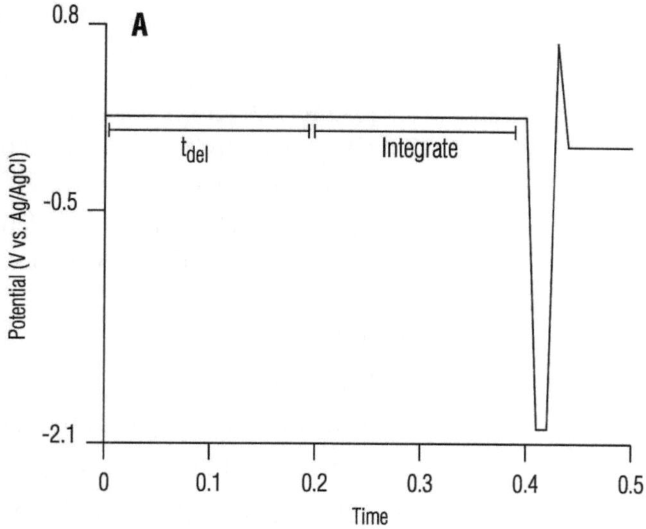


**Fig. 1** Waveform A for carbohydrate analysis

(Ag/AgCl). This leads to the oxidation of carbohydrates at the surface of the gold electrode. The current measure is proportional to the concentration of the carbohydrate; thus, quantifying using calibration curves with an external standard method is uncomplicated. A major drawback is that the oxidation products poison the surface of the electrode. A series of potentials is applied for fixed time periods after the detection potential to clean the surface. This series of potentials is called a waveform. Repeated application of a waveform is the basis of pulsed amperometry (Fig. 1) [10].

The HPAEC-PAD is extremely selective and specific for carbohydrate analysis, including the quantification of monosaccharides, disaccharides, oligosaccharides, sugar acids, and sugar alcohols. Pulsed amperometry only detects compounds that contain functional groups that are oxidizable at the detection voltage employed. Neutral or cationic sample components in the matrix elute in, or close to, the void volume of the column. In comparison to refractive index detection, there are no background peaks caused by media components or stress-inducing components, such as sodium chloride. Furthermore, the sensitivity of this method is outstanding and gives us a great advantage in detecting osmolytes in small extraction quantities [10].

## 2 Materials

### *2.1 Cultivation*

1. Erlenmeyer flasks with one baffle.
2. CMG: complete medium with 2% glucose. To make 1 L, mix 20 g glucose, 1 g yeast extract, 2 g peptone, 1 g casamino acid, 50 mL nitrate salt solution (containing 120 g $NaNO_3$, 10.4 g

KCl, 30.4 g $KH_2PO_4$, 10.4 g $MgSO_4 \times 7H_2O$ per liter $H_2O$), and 1 mL of a trace element solution (containing 22 g $ZnSO_4 \times 7H_2O$, 11 g $H_3BO_3$, 5 g $MnCl_2 \times 4H_2O$, 5 g $FeSO_4 \times 7H_2O$, 1.7 g $CoCl_2 \times 6H_2O$, 1.6 g $CuSO_4 \times 5H_2O$, 1.5 g $Na_2MoO_4 \times 2H_2O$, 50 g $Na_2EDTA$ per liter $H_2O$, pH 6.5 adjusted by 1 M KOH) [11].

3. Induction (stress) medium (depending on research of interest, e.g., sterile 50 mL 2.5 M KCl stock solution for 250 mL liquid culture).
4. Autoclave.
5. Flask shaker.
6. Sampling tubes (50 mL) (e.g., centrifuge tubes).
7. Duran glass filtering flask equipped with a Buchner funnel.
8. Filter papers (e.g., type: MN 615).
9. Vacuum pump or aspirator pump.
10. Deionized water.
11. Spatula.
12. Weighted plastic tubes (e.g., 15 mL centrifuge tube) (*see* **Note 1**).
13. Plastic tube caps with and without hole (*see* **Note 1**).
14. Liquid nitrogen.
15. Deep freezer (−80 °C).

### 2.2 Extraction

1. Lyophilizing equipment.
2. Lyophilized mycelium samples in weighted tubes.
3. Methanol.
4. Dichloromethane.
5. Thermoshaker for tubes (37 °C).
6. Scintillation vials (e.g., from Carl Roth, Germany).
7. Speed vacuum concentrator.

### 2.3 HPLC Standards

1. Carbohydrate standards in high quality (e.g., glycerol p.a., erythritol, D(+)-arabitol, D-sorbitol, D(+)-trehalose dihydrate, D(−)-mannitol, glucose monohydrate).
2. Analytical balance.
3. Double distilled water (18.2 MΩ).
4. Ultrasonic bath (e.g., from Bandelin).
5. Syringe without needle, sterile, 1 mL (e.g., from Terumo, Germany).
6. Syringe filters (e.g., from Macherey-Nagel) (*see* **Note 14**).

### 2.4 Sample Preparation

1. Lyophilized sample.
2. Double distilled water (18.2 MΩ).
3. Ultrasonic bath (e.g., from Bandelin).
4. Syringe without needle, sterile, 1 mL (e.g., from Terumo, Germany).
5. Syringe filters (e.g., from Macherey-Nagel) (*see* **Note 14**).

### 2.5 Eluent Preparation

1. 50% w/w aqueous sodium hydroxide.
2. Double distilled water (18.2 MΩ).
3. Vacuum filtration unit (e.g., Sartorius).
4. Filter papers.
5. Exsiccator.
6. Pipette.

### 2.6 HPAEC-PAD Analysis

1. Shimadzu LC system equipped with two LC-10Ai pumps, a DGU-20A degassing unit, a SIL-10Ai autosampler, a CBM-20A controller, and a CTO-20 AC column oven coupled to a Dionex ED40 Electrochemical Detector used in pulsed amperometric mode (*see* **Notes 13** and **15**).
2. An anion-exchange column of CarboPac MA1 (4 × 250 mm) combined with a guard column of CarboPac MA1 (4 × 50 mm) at 20 °C.
3. Software LabSolution Vers. 5.92 for data acquisition and processing.
4. Autosampler vials.
5. Eluent, prepare 1 L of a 480 mM NaOH solution by taking out 24.96 mL of the degassed double distilled water and adding the same volume of a 50% NaOH solution (*see* **Note 11**).
6. Double distilled water, 18.2 MΩ used for all eluent and standard preparations.

## 3 Methods

### 3.1 Cultivation of Fungal Cultures

Scale the fermentation to a volume of 250 mL liquid culture in Erlenmeyer flasks to enable parallel cultivation and the investigation of multiple *M. oryzae* strains while still ensuring robust handling and reproducibility.

1. Inoculate 250 mL liquid CMG with *M. oryzae* (three agar pieces of ca. 1 $cm^2$ with mycelium) and cultivate in Erlenmeyer flasks with one baffle at 26 °C and 100 rpm. (*see* **Note 3**).

2. Take sample zero (50 mL) 72 h after inoculation (corresponds to 20 mg mycelium, depending on the individual growth speed of the *M. oryzae* strain).
3. Separate mycelium of the 50 mL sample from culture broth by filtration using a glass filtering flask equipped with a Buchner funnel with filter paper and a vacuum pump.
4. Flush mycelium on filter paper with 3 volume units of $H_2O$ (150 mL) rapidly.
5. Transfer washed mycelium into plastic tube immediately by using a spatula and shock-freeze in liquid nitrogen. Store in deep freezer (−80 °C) until lyophilization starts. Avoid thawing of shock-frozen mycelium before lyophilization.
6. Then, induce cultivated strains with fivefold higher concentrated stressors/compounds dissolved in a volume of 50 mL $H_2O$. Subsequently, take samples at the time points to be investigated (*see* **Note 4**).

### 3.2 Extraction

1. Lyophilize frozen mycelium. Store in freezer until extraction starts.
2. Remove cap with hole from tube with lyophilized mycelium inside and replace with standard cap after adding 4 mL solvent (12:5 methanol/dichloromethane) per sample (*see* **Note 5**).
3. Place samples in a thermoshaker at 37 °C for 30 min.
4. Centrifuge samples for 5 min at 2900 × *g* (*see* **Note 6**).
5. Transfer approximately 3 mL of the supernatant into scintillation vials (*see* **Note 2**), for example, by decanting, but avoid transferring mycelium parts into the scintillation vials (*see* **Note 7**).
6. Re-extraction (*see* **Note 8**): Add 3 mL solvent (mix 12:5 methanol/dichloromethane) per sample and vortex the tube or shake vigorously to loosen the pelleted mycelium (*see* **Note 5**).
7. Repeat **step 3**: Place samples in a thermoshaker at 37 °C for 30 min.
8. Centrifuge samples for >10 min at 2900 × *g* (*see* **Note 9**). Decant, but avoid transferring mycelium parts into the scintillation vials (*see* **Note 7**).
9. Repeat **step 5**: Transfer supernatant into scintillation vials (*see* **Note 2**), for example, by decanting, but avoid transferring mycelium parts into the scintillation vials (*see* **Note 7**).
10. Evaporate solvents in a vacuum concentrator for approximately 2–3 h until the extracts are completely dry. Store extracts at −20 °C (!) or go on with sample preparation for HPAEC-PAD (Subheading 3.6) (*see* **Note 10**).

### 3.3 HPLC Standards

1. Weigh approximately 10 mg of each carbohydrate standard into a small beaker.
2. Notice the *exact* weight.
3. Dissolve in a small amount of double distilled water.
4. Transfer the solution to a volumetric flask and fill up to 100 mL with double distilled water.
5. Filter the standard solution using a sterile 0.5µm filter unit (e.g., from StarLab).
6. This stock solution can be stored for 4 weeks at 4 °C.
7. Withdraw 500µL of the stock solution under sterile conditions and dilute the solution 1:2 with double distilled water for the HPLC analysis. This standard solution should be prepared freshly once a day.
8. Transfer it to an autosampler vial.

### 3.4 Sample Preparation

1. Add 1 mL double distilled water to the lyophilized extract.
2. Dissolve it by sonification.
3. Filter through a syringe filter (pore size of 0.2µm) (*see* **Note 16**).
4. Transfer the sample into an autosampler vial.
5. Dilution with double distilled water may be necessary depending on the extract weight and the sensitivity of the detector.

### 3.5 Eluent Preparation

1. Filter 1 L of double distilled water through a 0.2µm filter using a vacuum filtering unit.
2. Degas the water in an exsiccator for 30 min (*see* **Notes 9** and **12**).
3. Prepare a 480 mM NaOH solution by taking out 24.96 mL of the degassed double distilled water and adding the same volume of a 50% NaOH solution (*see* **Note 11**).
4. Mix gently [12].

### 3.6 HPAEC-PAD Analysis

#### 3.6.1 Method Parameters

| | |
|---|---|
| Column | CarboPac MA1 Analytical (4 × 250 mm) with a CarboPac MA1 Guard Column (4 × 50 mm) |
| Temperature | 20 °C |
| Injection volume | 10 µL |
| Eluent | 480 mM NaOH |
| Flow rate | 0.4 mL/min, isocratic |
| Analysis time | 70 min |
| Detection | Pulsed amperometry, disposable gold working electrode, Ag/AgCl reference electrode |

| Waveform | Time (s) | Potential (V) | Integration |
|---|---|---|---|
| | 0.00 | +0.1 | |
| | 0.20 | +0.1 | Begin |
| | 0.40 | +0.1 | End |
| | 0.41 | −2.0 | |
| | 0.42 | −2.0 | |
| | 0.43 | +0.6 | |
| | 0.44 | −0.1 | |
| | 0.50 | −0.1 | |

#### 3.6.2 Qualitative and Quantitative Analysis

1. Sample components are identified by comparison with the retention times of the standards.
2. Quantification is based on the external standard method by using a three-level calibration curve.
3. Automatic quantification is performed by the software LabSolution Vers. 5.92 (Fa. Shimadzu).

   An example of a separation of a standard mixture of carbohydrates is given in Fig. 2.

## 4 Notes

1. Weigh the empty tubes without the cap, because two different caps per tube are used in this method: one with an intact cap and one with a hole in the cap (for better handling during lyophilization); use scissors, for example, to manually prepare caps with holes.
2. It is better to use glass than opaque plastic vials to find the extract pellet more easily for dissolving in water before the HPAEC-PAD analysis; use vials with at least 20 mL volume to avoid extract loss due to foaming and boiling delay of the solvents during evaporation.
3. Treatments with different stimuli should be carried out in the exponential growth phase of *M. oryzae* for reproducible results. It was found that after complete consumption of the free glucose in the medium, almost exclusively trehalose was enriched intracellularly and that the intracellular concentrations of the carbohydrates did not change strongly even after osmotic shock, not like that described by the pioneers Dixon et al. [3]. Thus, the experiment has to be terminated before the free glucose content in the medium is completely consumed by the fungus. Enough mycelium has to be available for each

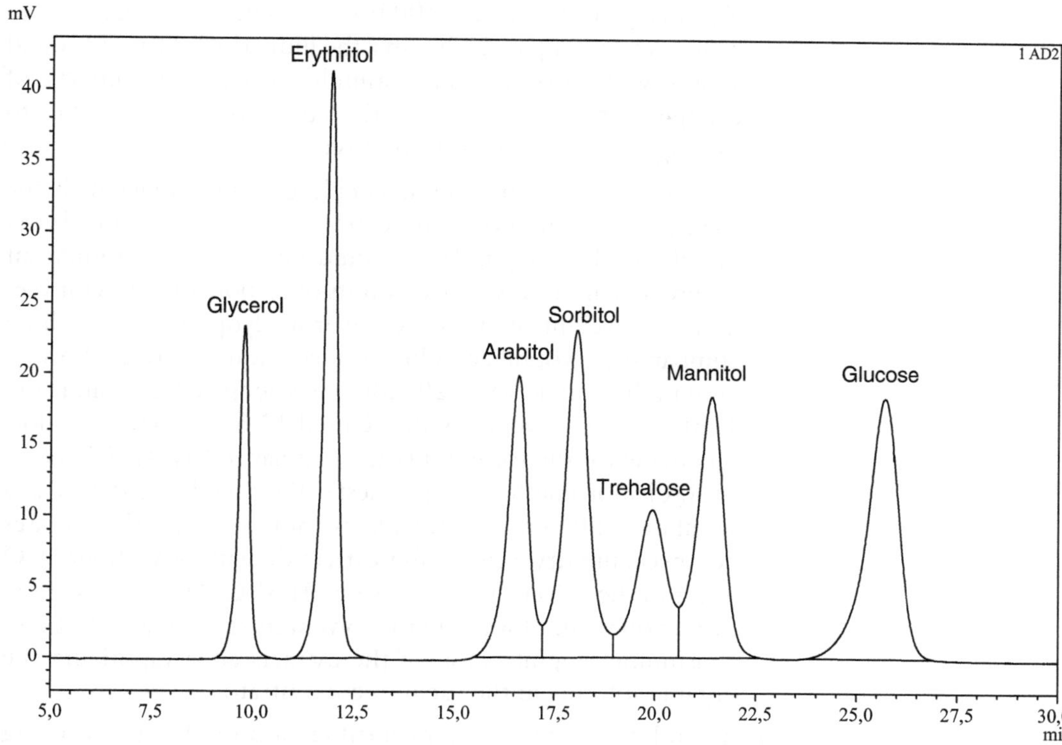


**Fig. 2** Separation of a standard mixture of carbohydrates: glycerol 27.5 mg/L (RT 9.781 min), erythritol 49.0 mg/L (RT 11.870 min), arabitol 51.5 mg/L (RT 16.440 min), sorbitol 76.5 mg/L (RT 17.864 min), trehalose 66.0 mg/L (RT 19.472 min), mannitol 67.0 mg/L (RT 21.067 min), glucose 82.0 mg/L (RT 24.938 min)

sample for a good and stable handling of the sample processing. Thus, the experiments can be started not earlier than 72 h after inoculation – at a time when the cultivated strains have already produced sufficient amounts of mycelium. We cultivated the strains in CM with 2% instead of 1% glucose, since the intracellular carbohydrate concentrations 24 h after stimulus is of interest in a lot of our experiments. Thus, the free glucose in the medium is not consumed before ca. 100–120 h after inoculation, ensuring that sufficient free glucose was available at all sampling time points of the experiment (check glucose level every day in your experimental setup).

4. The induction of the cultivated strains is carried out with substances or compounds in fivefold increased concentration, dissolved in a volume of sterile 50 mL $H_2O$, so that a onefold concentration of the corresponding stimulus in the Erlenmeyer flask is achieved after adding the 50 mL to the 200 mL culture (250 mL starting culture with 50 mL sampling subtracted before induction) in the Erlenmeyer flask at time point $t = 0$. The cultures are treated with 50 mL $H_2O$ for the negative control. Up to 50 mL of each flask is sampled per time point

(optimally not more than 100 mL) immediately before (0 h) as well as *x* h (*x* = up to 24 h) after the induction of the cultivated strains with corresponding stimuli to avoid the effects of changing shearing forces on the cultured mycelium due to varying culture volumes in the flasks.

5. Ensure the whole mycelium sample gets in contact with the solvent by flicking down mycelium (try not to shred or break mycelium; do not grind); we could not detect any significant differences in the absolute amounts of carbohydrates per mycelium dry weight after extraction from lyophilized mycelium compared to lyophilized plus grinded (liquid nitrogen) mycelium in the sample sizes (20–200 mg lyophilized mycelium and 4 mL + 3 mL solvent) used. We used 12:5 methanol/dichloromethane for the solvent mixture (similar to [3]: 12:5:3 methanol/dichloromethane/aqua dest). We could not detect any significant differences in the relative amounts of carbohydrates per mycelium dry weight after extraction with or without $H_2O$ water in the solvent mixture using HPAEC-PAD; we used the extraction without water in the solvent mixture to avoid a time-consuming lyophilization of the extracts after vaporizing the organic solvents in the vacuum concentrator.
6. Avoid longer and stronger centrifugation resulting in a strong sticking of the pellet at the bottom of the tube.
7. You do not need to transfer the whole volume of supernatant into the scintillation vial at this step, because the pelleted mycelium will be re-extracted (*see* Subheading 3.2, **step 6**). Depending on your scintillation vial size (*see* also **Note 2**), you can already start to vacuum concentrate the dissolved extracts (*see* Subheading 3.2, **step 10**) and, thus, reduce the solvent volume in the vial to avoid boiling delay caused by too much liquid in the vial after transferring the supernatant from the re-extraction step (3 + 4 mL).
8. Re-extraction of the same mycelium samples increases the amount of carbohydrate per mycelium dry weight by approximately 20%. A re-re-extraction of the same mycelium samples did not increase the amount of carbohydrate per mycelium dry weight significantly.
9. A longer and stronger centrifugation will facilitate the clean transfer (via decanting) of the supernatant into the scintillation vial without transferring mycelium parts.
10. Avoid boiling delay (*see* also **Note 2**), depending on the cultivation conditions or treatment of the *M. oryzae* culture. Extracts may be oily, so complete dryness will not be reached; weigh scintillation vials (without caps) before and after extraction if the mycelium/extract dry weight ratio is important for you.

11. An accurate preparation and use of the eluents is very important for a successful analysis. The eluent consists of two components: water and sodium hydroxide. The water used should be high-quality double distilled water of low resistivity (18 MΩ). It is filtered through a 0.2μm filter to remove particles and possible biological contaminants. Water has to be degassed under vacuum for 30 min before adding the NaOH solution [4].
12. It is very important to avoid carbonate contamination. Carbon dioxide dissolved in water converts to carbonate at pH $\geq$ 12. Carbonate is a divalent anion that is a stronger eluent than hydroxide. Excessive carbonate in eluents causes a reduction of carbohydrate retention and resolution [4].
13. A 50% (w/w) sodium hydroxide solution should be used for the eluent preparation. *Never use sodium hydroxide pellets, because they are covered with a thin layer of sodium carbonate.* Remove the sodium hydroxide solution from the middle of the bottle with a plastic pipette and do not use the solution at the bottom to avoid carbonate contamination. Gently swirl the bottle to complete the mixing [4].
14. Fresh eluent should be prepared daily or weekly in the case of using a helium degassing unit.
15. The whole HPLC system must be free of metal, including the eluent flow path and autosampler. Metal contamination of the analytical column leads to poor peak efficiency and reproducibility.
16. Pay attention to the filter materials because of glycerol and glucose contamination. Some filters are treated with glycerol and filters made of cellulose could falsify the results.

## References


1. Jackson TR, Springall JS, Rogalle D, Masumoto N, Ching Li H, D'Hooge F et al (2008) Boronate affinity saccharide electrophoresis: a novel carbohydrate analysis tool. Electrophoresis 29(20):4185–4191
2. Saito H, Tatebayashi K (2004) Regulation of the osmoregulatory HOG MAPK cascade in yeast. J Biochem 136(3):267–272
3. Dixon K, Xu J, Smirnoff N, Talbot N (1999) Independent signaling pathways regulate cellular turgor during hyperosmotic stress and appressorium-mediated plant infection by *Magnaporthe grisea*. Plant Cell 11 (10):2045–2058
4. Bohnert S, Antelo L, Grünewald C, Yemelin A, Andresen K, Jacob S (2019) Rapid adaptation of signaling networks in the fungal pathogen *Magnaporthe oryzae*. BMC Genomics 20:763
5. Gil Ortiz R, Lull Noguera C, Boscaiu Neagu MT, Bautista Carrascosa I, Lidón Cerezuela AL, Vicente Meana Ó (2011) Soluble carbohydrates as osmolytes in several halophytes from a mediterranean salt marsh. Notulae Botanicae Horti Agrobotanici Cluj-Napoca 39(2):9–17
6. Dionex Application Note 122: "The Determination of Carbohydrates, Alcohols, and Glycols in Fermentation Broths"
7. Dionex Application Note 87: "Determination of Sugar Alcohols in Confections and Fruit Juices by High-Performance Anion-Exchange Chromatography with Pulsed Amperometric Detection"

8. Dionex Application Note 92: "The Determination of Sugars in Molasses by High-Performance Anion Exchange with Pulsed Amperometric Detection"
9. Dionex Technical Note 20: "Analysis of Carbohydrates by Anion Exchange Chromatography with Pulsed Amperometric Detection"
10. Dionex Technical Note 21: "Optimal Settings for Pulsed Amperometric Detection of Carbohydrates Using the Dionex ED40 Electrochemical Detector"
11. Jacob S, Foster AJ, Yemelin A, Thines E (2014) Histidine kinases mediate differentiation, stress response, and pathogenicity in *Magnaporthe oryzae*. MicrobiologyOpen 3(5):668–687
12. Dionex Technical Note 71: "Eluent Preparation for High-Performance Anion-Exchange Chromatography with Pulsed Amperometric Detection"

# Chapter 5

# Characterizing the Secretion Systems of *Magnaporthe oryzae*

**Ely Oliveira-Garcia and Barbara Valent**


## Abstract

Pharmacological approaches have made a tremendous impact on the field of microbial secretion systems. This protocol describes the inhibition of Golgi-dependent secretion in *Magnaporthe oryzae* though brefeldin A (BFA) treatment. State-of-the-art live-cell imaging allows tracking secreted proteins in their secretion pathways. Here we applied this protocol for defining the secretion systems of two fluorescently labeled effectors, Bas4 (apoplastic) and Pwl2 (cytoplasmic). Secretion of Bas4 is clearly inhibited by brefeldin A (BFA), indicating its Golgi-dependent secretion pathway. By contrast, secretion of Pwl2 is BFA insensitive and follows a nonconventional secretion pathway that is Snare and Exocyst dependent. The protocol is suitable to other plant-microbial systems and in vitro secreted microbial proteins.


**Key words** Secreted fungal proteins, Pharmacological approach, Chemical inhibition, Golgi-dependent secretion, Snare- and Exocyst-dependent secretion, Effectors, Fluorescently labeled proteins, In planta secretion

## 1 Introduction

Microbial secreted proteins merit detailed consideration because they play many essential roles for adaptation to their environment and host colonization [1–3]. Biotrophic microbial pathogens grow in their host without damaging the living host cells [4, 5] and secrete effector proteins into and around the host cells [6, 7]. Bacterial effectors have multiple important functions, including induction of the host cell entry by intracellular pathogens, reprogram of host cell signaling, and suppression of the immune responses. These specific functions are associated with distinct forms of secretion that function differently during pathogenesis [2, 8–11]. Eukaryotic filamentous pathogens, fungi and oomycetes, secrete an even larger number of effector proteins during their biotrophic development. Functional characterization of these effectors revealed their roles in suppressing plant immunity [12–14], reprograming metabolism [15], and evading recognition of the

Stefan Jacob (ed.), *Magnaporthe oryzae: Methods and Protocols*, Methods in Molecular Biology, vol. 2356,
https://doi.org/10.1007/978-1-0716-1613-0_5, © Springer Science+Business Media, LLC, part of Springer Nature 2021

invading microbe [16, 17]. Much of our current knowledge on effector functionality comes from targeted mutagenesis and live-cell imaging studies. Live-cell imaging studies have clearly shown that fungal effectors are delivered both inside of the host cells and to the host-pathogen interface during infection [6, 18]. To cause disease, the hemibiotrophic ascomycete fungus *Magnaporthe oryzae* expresses many low-molecular-weight biotrophy-associated secreted (Bas) proteins, including known effector proteins, and these proteins possess classical signal peptides, which facilitate delivery into the endoplasmic reticulum (ER) [6, 11, 19]. Likewise, the ER chaperone, Lhs1, is essential for secretion of effectors [20]. Live-cell imaging of *M. oryzae* strains secreting chimeric effector proteins labeled with carboxyterminal green fluorescent protein (GFP), monomeric red fluorescent protein (mRFP), or monomeric Cherry (mCherry) facilitates the prediction of effector functions in planta [6, 18, 21]. Effectors are also classified by their patterns of accumulation relative to the EIHM (extrainvasive hyphal membrane), the altered plant membrane that the fungus uses for establishing the biotrophic interaction. Remarkably, apoplastic effectors do not cross the EIHM, residing in the extracellular plant compartment, the biotrophic interphase [3, 18, 22, 23]. However, cytoplasmic effectors are translocated into the cyto plasm of living plant cells [24]. These accumulation patterns suggested the probable distinct secretion pathways by which *M. oryzae* secretes apoplastic and cytoplasmic effectors. Indeed, pharmacological approaches confirmed this interesting hypothesis [22]. Secretion of apoplast effectors is brefeldin A (BFA) sensitive, indicating their conventional Golgi-dependent secretion. By contrast, secretion of cytoplasmic effectors is BFA insensitive, being secreted by a nonconventional, Golgi-independent secretion system into a specific extrainvasive hyphal matrix (EIHMx) region called the biotrophic interfacial complex (BIC) [22]. Furthermore, target mutagenesis approaches revealed that the system employed by *M. oryzae* secreting cytoplasmic effectors is Snare and Exocyst dependent [3, 22]. These findings may be relevant to other biotrophic filamentous microorganism systems; therefore, we are sharing the protocol for BFA treatment for inhibition of secreted proteins in planta. In this chapter, we describe the detailed protocol for pharmacology treatment of leaf sheaths infected by *M. oryzae* that includes infecting and trimming leaf sheaths, inhibitor treatment, as well as analyses under microscope.

## 2 Materials

Prepare all solutions using double distilled water or ultrapure water (prepared by purifying deionized water to attain a sensitivity of 18 M cm at 25 °C) and analytical grade reagents. Prepare and store all reagents at room temperature (unless indicated otherwise). Diligently follow the waste disposal regulation when disposing waste materials. Inactivate genetically modified microbial strains and infected tissue by autoclaving in liquid cycle [121 °C (250 °F), 15 psi] for at least 30 min. Use 70% ethanol or CaviCide for surface decontamination.

1. Brefeldin A (BFA) stock—brefeldin A in DMSO 100%.
2. 0.1% DMSO (dimethyl sulfoxide)—prepare 0.1% DMSO from 100% DMSO by a 1000-fold dilution: for instance, 900 μL in a total 900 mL (i.e., 900 μL added to 899.1 mL double distilled water) (*see* **Note 1**).
3. Sterile 0.25% gelatin solution (gelatin type B: from bovine skin).
4. Incubate the rice seedlings in growth chamber at 25 °C (12-h light and 12-h dark) to the V6–V7 growth stage (collar formation on leaves 6–7 on the main stem). The growth stages V6 and V7 are generally obtained in 3 to 4 weeks under these conditions (*see* **Note 2**).
5. Fungal stocks are obtained from filter paper disks containing fungal mycelia. Stocks must be completely dehydrated in desiccator for 2 weeks and stored in dry sterile tubes or envelopes at −80 °C or −20 °C (*see* **Note 3**).
6. Rice bran agar (RBA). Transfer 25 g of rice bran (product of polishing rice seeds) and 15 g of bacto-agar to a 2 L Erlenmeyer. Bring the volume up to 1 L with double distilled water. Autoclave for 25 min. Pour the warm media (approximately 45–55 °C) in sterile Petri dishes (100 × 15 mm). Store in room temperature (23–25 °C) or at 4 °C. The rice bran agar plates should be thickly poured.
7. Oatmeal agar (OMA). Heat 50 g of old fashion oats and 15 g of bacto-agar at 70 °C in 500 mL of double distilled water for 40 min. Filter the solution in a double layer of cheesecloth, squeezing until most of the liquid is removed from the remaining oat solids. Transfer the suspension that passed through the filter into a 2-L Erlenmeyer and bring the volume up to 1 L with double distilled water. Autoclave for 45 min. Pour the warm media (approximately 55–45 °C) in sterile Petri dishes (100 × 15 mm). Store in room temperature (23–25 °C) or at 4 °C. The oatmeal agar plates should be thickly poured (*see* **Note 4**).
8. Moist chamber. The moist chamber is composed of a sealable container with 2–3 layers of moist paper towels. Add at least two PCR tube strips (Fig. 1) for holding the rice sheaths and avoiding contact with the paper towels and possible contaminates.

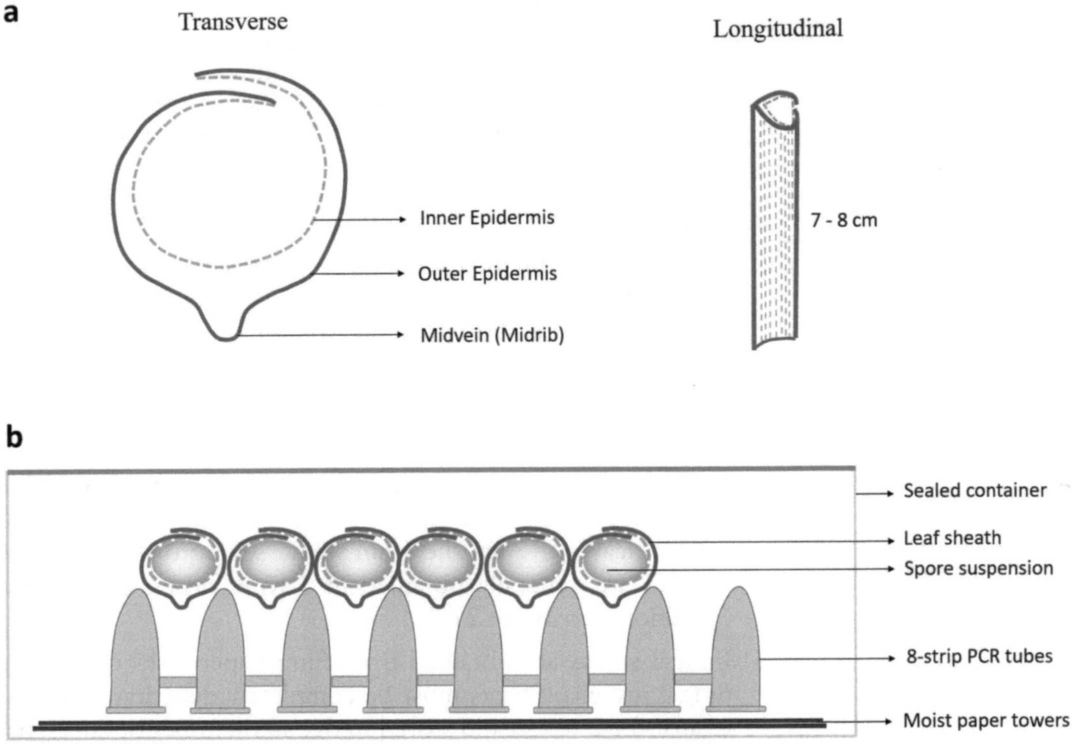


**Fig. 1** Leaf sheath inoculation assay. (**a**) These illustrations show transverse and longitudinal views of the rice leaf sheath used for the infection assay. The dashed line represents the inner epidermis; solid line is the outer epidermis; and the convex structure represents the sheath midvein or midrib. (**b**) Moist chamber used for infecting the rice leaf sheath with *Magnaporthe oryzae*. The 7–8 cm long leaf sheaths are mounted on the inverted eight-strip PCR tubes in a sealable moist chamber containing wet paper towels. The spore suspension is applied into the extremity of the sheath until completely filled. The moist chamber is sealed and incubated at 25 °C in darkness for at least 24 h

## 3 Methods

Carry out all procedures at room temperature unless otherwise specified.

### 3.1 Fungal Culture, Spore Suspension Preparation, and Leaf Sheath Inoculation

Begin growing the fungus about 7–10 days before the start of the rice blast assay (*see* **Note 5**).

1. Start the fungal cultures from the stock filter papers stored at −20 °C (in the hood under sterile conditions). Cut a small piece of the filter paper onto a rice bran or oatmeal agar plate (100 × 15 mm Petri dish). Alternatively transfer a small plug from a plate that is growing in the incubator, that is less than 14 days old, and that was originally inoculated with filter paper. *M. oryzae* strains carrying the fluorescently labeled secreted

proteins should be grown on RBA or OMA at 25 °C under constant fluorescent lights for 4–5 days (*see* **Note 6**).

2. Harvest the asexual spores from the medium surface by adding 5 mL of sterile 0.25% gelatin.
3. Mix gently the spores with the gelatin with a sterile pipette tip or a glass spreader rod.
4. Filter the spore suspension through one layer of sterile Miracloth.
5. Transfer the suspension to a 15 mL Falcon tube and centrifuge for 5000 rpm for 3 min.
6. Discard supernatant and resuspend the spore pellet in new gelatin solution.
7. Adjust the spore concentration to 50,000 spores/mL using a hemocytometer.

   For each fungal strain, put a drop of each 50,000 spore/mL solution onto a YEG plate (2 g yeast extract, 10 g glucose, 17 g agar per 1 L). This is so—that you can check the germination rate the next day (*see* **Note 7**).

### 3.2 Inoculation Assay of Rice Leaf Sheaths

Obtain leaf sheaths from the second youngest leaf of the main stem of a rice plant.

1. Cut 7–8 cm long leaf sheaths and add on the interval of each two tubes of the inverted eight-strip PCR tubes in the moist chamber (Fig. 2a).
2. Apply the spore suspension in the extremity of the sheath with a micropipette and sterile 1 mL tips until the entire interior of the leaf sheath is filled (Fig. 2b). Prepare at least three non-inoculated sheaths as controls (mock control).
3. Seal the moist chamber and incubate in the dark at 25 °C for at least 24 h.
4. After the incubation period, the sheaths are ready for the next step, sectioning.

### 3.3 Leaf Sheath Sections of Rice

The efficiency of the pharmacological assay is directly associated with the quality and thickness of the leaf sheath sections. Obtain ultrathin sections (3–4 layers of plant cells) with sharp blades in a three-step sectioning procedure (Fig. 2).

The steps include:

1. Cut the sides of the leaf sheath.
2. Cut the midvein or midrib.
3. Holding the section on your fingers, try to cut as much outer epidermal layer as you can. The sections are ready to mount on the slide for microscopy (*see* **Notes 8** and **9**).

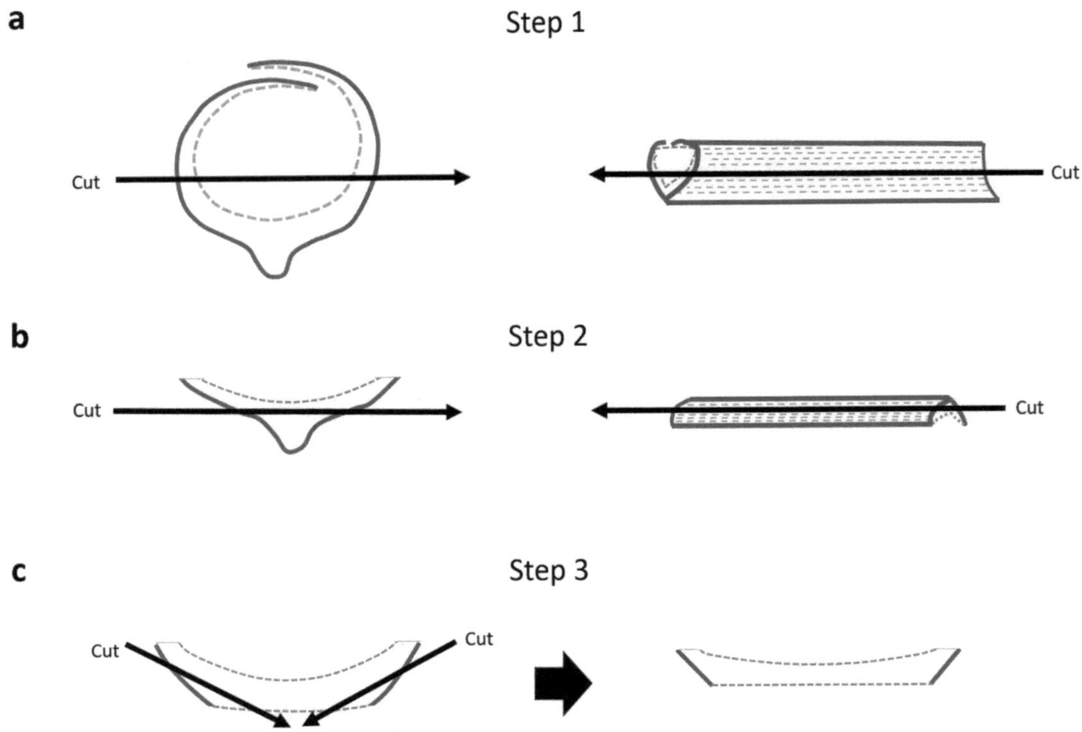


**Fig. 2** Infected leaf sheath trimming for pharmacological treatment and live-cell imaging studies of fungal secreted proteins. The three-step protocol for obtaining sheath sections suitable to pharmacological treatment involves (**a**) leaf sheath side trimming/removal, (**b**) leaf sheath midvein trimming/removal, and (**c**) leaf sheath outer epidermis trimming/removal. The product is ready for the 3-h chemical treatment and mounting on the slide for microscopy. The transverse section views are shown in (**a**), (**b**) (left pictures), and (**c**). The longitudinal sections are shown in (**a**) and (**b**) (left pictures)

### 3.4 Treatment of Infected Tissue with Brefeldin A (BFA)

1. Cut the leaf sheath into sections of 1 cm long.
2. Place the section in microcentrifuge tubes with enough 50 μg/mL BFA solution to cover the sheaths.
3. Treat at least three sheaths with 0.1% DMSO or corresponding DMSO concentration of the targeted chemical applied.
4. After a 3-h incubation at room temperature, the sheaths are ready for microscopy. BFA treatment blocks Golgi-dependent secretion. Observe the positive control Bas4:eGFP under fluorescence microscope with the appropriate filter sets.

   Whereas in the DMSO control, Bas4 is secreted in the apoplast (Fig. 3), treatment with BFA leads to a massive accumulation of Bas4:eGFP in the fungal cytoplasm. BFA treatment does not affect secretion of Pwl:mRFP into the BIC (Fig. 3) and translocation into the rice cell cytoplasm.

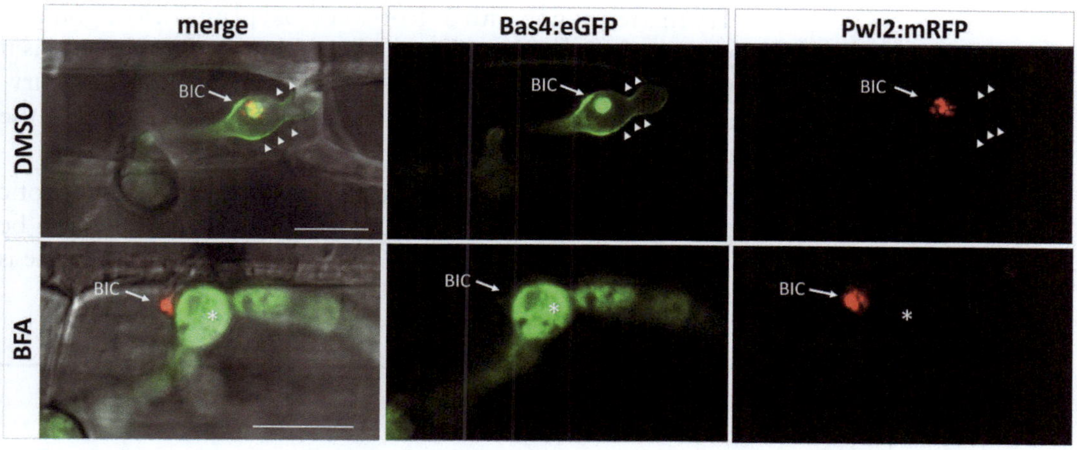


**Fig. 3** Brefeldin A blocks the secretion of apoplastic effector Bas4 but not of cytoplasmic effector, Pwl2. Images from left to right: either merged bright-field, eGFP, and mRFP; eGFP alone; or mRFP alone. After secretion, Bas4:eGFP (green) shows apoplastic localization outlining the invasive hypha (arrowheads). The cytoplasmic effector Pwl2:mRFP (red) shows preferential BIC localization (arrow) and translocation into the rice cell. In the presence of BFA, Bas4:eGFP (green) is retained in the fungal ER (asterisk), but Pwl2:mRFP remains BIC-localized (arrow), imaged with the same transformant at 3 h after exposure to BFA. Bars are 10 μm. *BFA* Brefeldin A, *DMSO* DMSO control

## 4 Notes

1. DMSO has a freezing point of 18.5 °C and will be frozen at 4 °C, which is the temperature at which the BFA should be stored.
2. Rice growth speed may vary due to many factors, including rice cultivar, soil properties, accuracy of the growth chamber settings, etc. Plants should grow up to 6 weeks. Avoid using old and unhealthy tissue for best results.
3. 10–20% glycerol stocks are suitable for many other genera of fungi.
4. Many fungi show satisfactory growth rates on OMA medium, which often induces sporulation.
5. This fungus grows slowly and it takes about 17 days to fill a 100 × 15 mm Petri dish with OMA or RBA.
6. *M. oryzae* is not stable for multiple transfers, showing spontaneous mutations after many subcultures. Therefore, the process of subculturing this fungus should be avoided.
7. Young spores show germination rates of approximately 95–98%. Plates that are 17 days or older generally show reduced germination rates and reduced appressorium penetration. Therefore, it is best to use 7–10-day-old plates, on which the fungus has not filled the entire Petri dish.

8. Try this technique with 3- to 4-week-old plants before going to younger leaf sheaths. Change blade for every 3 to 4 sections if you want a very thin section. Avoid extending, folding, curving, or any other procedures that damage the integrity of the plant cells.
9. This method is adaptable to other plant-filamentous eukaryotic pathogen systems. In the case of the maize, the tissue can be obtained from midvein leaf of seedlings. The trimmed tissue is suitable for live-cell imaging [5].

## Acknowledgments

This work was supported by the Louisiana State University AgCenter HATCH and the NIFA-USDA grant (#2017-67013-26525).

### References

1. Doehlemann G, Wahl R, Horst RJ, Voll LM, Usadel B, Poree F, Stitt M, Pons-Kühnemann J, Sonnewald U, Kahmann R et al (2008) Reprogramming a maize plant: transcriptional and metabolic changes induced by the fungal biotroph *Ustilago maydis*. Plant J 56:181–195
2. Schirawski J, Mannhaupt G, Münch K, Brefort T, Schipper K, Doehlemann G, Di Stasio M, Rössel N, Mendoza-Mendoza A, Pester D et al (2010) Pathogenicity determinants in smut fungi revealed by genome comparison. Science 330:1546–1548
3. Oliveira-Garcia E, Valent B (2015) How eukaryotic filamentous pathogens evade plant recognition. Curr Opin Microbiol 26:92–101
4. Kankanala P, Czymmek K, Valent B (2007) Roles for rice membrane dynamics and plasmodesmata during biotrophic invasion by the blast fungus. Plant Cell 19:706–724
5. Oliveira-Garcia E, Deising HB (2013) Infection structure-specific expression of β-1,3-glucan synthase is essential for pathogenicity of *Colletotrichum graminicola* and evasion of β-glucan-triggered immunity in maize. Plant Cell 25:2356–2378
6. Mosquera G, Giraldo MC, Khang CH, Coughlan S, Valent B (2009) Interaction transcriptome analysis identifies *Magnaporthe oryzae* BAS1-4 as biotrophy-associated secreted proteins in rice blast disease. Plant Cell 21:1273–1290
7. Presti LL, Lanver D, Schweizer G, Tanaka S, Liang L, Tollot M, Zuccaro A, Reissmann S, Kahmann R (2015) Fungal effectors and plant susceptibility. Annu Rev Plant Biol 66:513–545
8. Alfano JR, Collmer A (2004) Type III secretion system effector proteins: double agents in bacterial disease and plant defense. Annu Rev Phytopathol 42:385–414
9. Galán JE, Collmer A (1999) Type III secretion machines: bacterial devices for protein delivery into host cells. Science 284:1322–1328
10. Rafiqi M, Ellis JG, Ludowici VA, Hardham AR, Dodds PN (2012) Challenges and progress towards understanding the role of effectors in plant–fungal interactions. Curr Opin Plant Biol 15:477–482
11. Valent B, Khang CH (2010) Recent advances in rice blast effector research. Curr Opin Plant Biol 13:434–441
12. Birch PRJ, Armstrong M, Bos J, Boevink P, Gilroy EM, Taylor RM, Wawra S, Pritchard L, Conti L, Ewan R et al (2009) Towards understanding the virulence functions of RXLR effectors of the oomycete plant pathogen *Phytophthora infestans*. J Exp Bot 60:1133–1140
13. Hogenhout SA, Van der Hoorn RAL, Terauchi R, Kamoun S (2009) Emerging concepts in effector biology of plant-associated organisms. Mol Plant Microbe Interact 22:115–122
14. Park CH, Chen S, Shirsekar G, Zhou B, Khang CH, Songkumarn P, Afzal AJ, Ning Y, Wang R, Bellizzi M et al (2012) The *Magnaporthe oryzae* effector AvrPiz-t targets the RING E3 ubiquitin ligase APIP6 to suppress pathogen-associated molecular pattern–triggered immunity in rice. Plant Cell 24:4748–4762

15. Djamei A, Schipper K, Rabe F, Ghosh A, Vincon V, Kahnt J, Osorio S, Tohge T, Fernie AR, Feussner I et al (2011) Metabolic priming by a secreted fungal effector. Nature 478:395–398
16. de Jonge R, Peter van Esse H, Kombrink A, Shinya T, Desaki Y, Bours R, van der Krol S, Shibuya N, Joosten MHAJ, Thomma BPHJ (2010) Conserved fungal LysM effector Ecp6 prevents chitin-triggered immunity in plants. Science 329:953–955
17. Mentlak TA, Kombrink A, Shinya T, Ryder LS, Otomo I, Saitoh H, Terauchi R, Nishizawa Y, Shibuya N, BPHJ T et al (2012) Effector-mediated suppression of chitin-triggered immunity by *Magnaporthe oryzae* is necessary for rice blast disease. Plant Cell 24:322–335
18. Khang CH, Berruyer R, Giraldo MC, Kankanala P, Park S-Y, Czymmek K, Kang S, Valent B (2010) Translocation of *Magnaporthe oryzae* effectors into rice cells and their subsequent cell-to-cell movement. Plant Cell 22:1388–1403
19. Saitoh H, Fujisawa S, Mitsuoka C, Ito A, Hirabuchi A, Ikeda K, Irieda H, Yoshino K, Yoshida K, Matsumura H et al (2012) Large-scale gene disruption in *Magnaporthe oryzae* identifies MC69, a secreted protein required for infection by monocot and dicot fungal pathogens. PLoS Pathog 8:e1002711
20. Yi M, Chi M-H, Khang CH, Park S-Y, Kang S, Valent B, Lee Y-H (2009) The ER chaperone LHS1 is involved in asexual development and rice infection by the blast fungus *Magnaporthe oryzae*. Plant Cell 21:681–695
21. Sakulkoo W, Osés-Ruiz M, Oliveira Garcia E, Soanes DM, Littlejohn GR, Hacker C, Correia A, Valent B, Talbot NJ (2018) A single fungal MAP kinase controls plant cell-to-cell invasion by the rice blast fungus. Science 359:1399–1403
22. Giraldo MC, Valent B (2013) Filamentous plant pathogen effectors in action. Nat Rev Microbiol 11:800
23. Yi M, Valent B (2013) Communication between filamentous pathogens and plants at the biotrophic interface. Annl Rev Phytopathol 51:587–611
24. Giraldo MC, Dagdas YF, Gupta YK, Mentlak TA, Yi M, Martinez-Rocha AL, Saitoh H, Terauchi R, Talbot NJ, Valent B (2013) Two distinct secretion systems facilitate tissue invasion by the rice blast fungus *Magnaporthe oryzae*. Nat Commun 4:1996–1996

# Chapter 6

# Specimen Preparation and Observations of *Magnaporthe oryzae* Appressorial Cells Under Electron Microscopy

**Raquel O. Rocha and Richard A. Wilson**

## Abstract

Electron microscopy (EM) allows characterization of the morphology and ultrastructure of a cell. However, challenges concerning cryo sample fixation are still one of the main roadblocks to its widespread adoption. In this protocol, we describe two alternative EM preparation methods employed to study *Magnaporthe oryzae* appressoria on artificial hydrophobic surfaces.

**Key words** Electron microscopy, Sample preparation, Surfaces, SEM, TEM, Imaging, Appressorium, *Magnaporthe oryzae*

## 1 Introduction

The rice blast pathosystem has emerged as a leading model for understanding plant-pathogen interactions due to advances in molecular genetics combined with cell imaging of both the pathogen and the host [1, 2]. The development of powerful new microscopic technologies and techniques, particularly those based on live-cell imaging and the use of fluorescent probes [3], has significantly enhanced our understanding of *Magnaporthe oryzae* physiology and infection-related development both on and in the host plant cells [4–10]. Incorporating new microscopic techniques into our workflow will continue to be critical to efforts aimed at decoding the links between the *M. oryzae* genome and cellular structure and functioning.

Among the leading technologies for cellular characterization, electron microscopy (EM) has grown to be an essential tool for characterizing the morphology and ultrastructure of a cell. By using accelerated free electrons, with shorter wavelengths than visible light, EM can generate better high-resolution images of smaller structures than light microscopy [11, 12]. Over time, access to electron instrumentation has increased to broader academic

Stefan Jacob (ed.), *Magnaporthe oryzae: Methods and Protocols*, Methods in Molecular Biology, vol. 2356, https://doi.org/10.1007/978-1-0716-1613-0_6, 

environments. Unfortunately, cryo sample fixation remains the biggest challenge to the use of electron microscopes. Fixation protocols relying on freezing samples are considered to be of the highest quality, generating fewer artifacts. However, specific equipment is required to prepare, maintain, and image cryofixated biological samples [13].

Here, we provide cost-effective alternative approaches to cryofixation that are suitable for rapidly preparing samples of *M. oryzae* appressoria for scanning electron microscopy (SEM) and transmission electron microscopy (TEM) analysis. Because appressorial morphogenesis from spores relies on sensing and attachment to hard hydrophobic surfaces, sample preparation of this particular stage needs to permit the removal of the appressorial structure from the hard surface while limiting alterations to its three-dimensional structure due to handling.

## 2 Materials

Prepare all solutions using ultrapure water and analytical grade reagents.

### 2.1 SEM Analysis

#### 2.1.1 Hydrophobic Surfaces for Appressoria Formation

1. Cellophane membranes (*see* **Note 1**).
2. Glass coverslips.

#### 2.1.2 SEM Sample Preparation

1. Water-agar medium, 3% agar (w/v) in water.
2. Bath sonicator.

### 2.2 TEM Analysis

#### 2.2.1 Hydrophobic Surfaces for Appressoria Formation

1. Glass coverslips.
2. Sof Sole® Water Proofer (*see* **Note 2**).

#### 2.2.2 TEM Sample Preparation

1. 4% agarose in water (melt agarose and keep it at 40 °C until use).
2. 2.5% glutaraldehyde fixative in water. Glutaraldehyde fixatives are prepared from EM grade 25% glutaraldehyde, found in sealed ampoules.
3. 2% osmium tetroxide in water. Prepared from EM grade osmium obtained in sealed glass ampoules.
4. 30%, 50%, 75%, 95%, and 100% ethanol in water.

5. Low Viscosity Embedding Media Spurr's Kit (Electron Microscopy Sciences, Inc., USA).
6. Leica EM UC7 Ultramicrotome.

## 3 Methods

Carry out all procedures at room temperature unless otherwise specified.

### *3.1 SEM Sample Preparation*

1. Suspend 3 g of agar in 100 mL distilled water. Heat to boiling to dissolve the agar completely. Sterilize the medium by autoclaving.
2. Place a layer (no more than 10 mm thick) of liquid water-agar on top of the glass coverslip. Wait until medium solidifies.
3. Gently lay the cellophane membrane on top of the water-agar (Fig. 1) (*see* **Note 1**). Make sure that the full extension of the membrane is in contact with the medium below.
4. Inoculate the spore suspension of *M. oryzae* ($1 \times 10^5$ spores/mL) on top of the cellophane membrane. Leave the spore-cellophane-agar-coverslip assembly supported in a Petri dish containing a wet filter paper. Allow the appressoria to form for 24 h in the dark at 24 °C.
5. After the incubation time, take the spore-cellophane assembly and sonicate it for 30 min (*see* **Note 3**). Be careful to maintain the orientation of the cellophane membrane with the appressoria uppermost.
6. After sonication, carefully transfer the cellophane membrane and place it on a Petri dish, the appressoria-containing surface facing up. Leave the membrane to air dry for up to 24 h.
7. Place samples on a double-sided adhesive conductive tab on an aluminum SEM sample-mounting stub.
8. Sputter coat samples with chromium (~5 nm thick) using a Denton Vacuum Desk V sputter coater. The samples are now ready for imaging (Fig. 2).

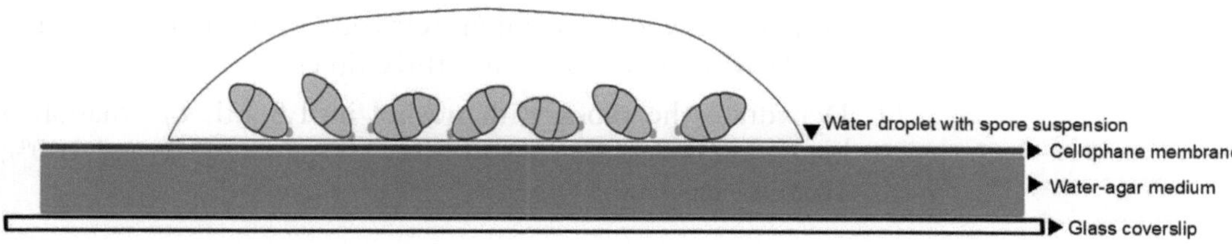


**Fig. 1** Schematic representation of sample assembly prior to SEM preparation and analysis. Spores of *M. oryzae* are inoculated on top of the cellophane membrane and allowed to form appressoria

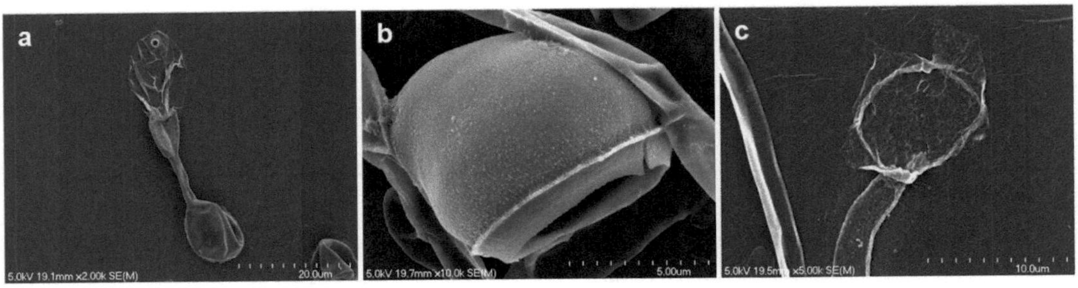


**Fig. 2** SEM images were taken after following the described protocol. (**a**) Spore and appressorium attached to the cellophane membrane. (**b**) Appressorium detached from the cellophane membrane with visible mucilage ring and unmelanized pore. (**c**) The imprint of an appressorium formed on the surface of cellophane. Note the attachment of both appressorium and germ tube

### 3.2 TEM Sample Preparation

1. Spray glass coverslips with Sof Sole® Water Proofer spray. Leave the coverslips to air dry for 48 h.
2. Inoculate the spore suspension of *M. oryzae* ($1 \times 10^5$ spores/mL) on the top of the coverslip. Leave the spore coverslip assembly supported in a Petri dish containing a wet filter paper. Allow the appressoria to form for 24 h in the dark at 24 °C.
3. After the incubation time, dry out the water droplet on the top of the coverslip with sterile filter paper. Dry the water droplet by its side to reduce interferences with appressoria surroundings.
4. Add 450 μL of the 2.5% glutaraldehyde fixative. Fix the samples for 1 h at room temperature (*see* **Note 4**).
5. Remove the glutaraldehyde solution by pipetting and drying out the remaining liquid with sterile filter paper.
6. Add 150 μL of 4% agarose (melted and at 40 °C) to the drop site. Wait until agarose is completely solidified.
7. Scrape off the coverslip using a sterile razor blade to remove appressoria enrobed with agarose (*see* **Note 5**).
8. Cut the agarose, as displayed in Fig. 3. Transfer appressoria-agarose blocks to 1.5 mL tube.
9. Add 1 mL of 2% osmium tetroxide fixative and incubate at room temperature for 2 h.
10. Properly discard the osmium tetroxide solution and wash the blocks with de-ionized water three times.
11. Dehydrate the block with washes in 1.5 mL of ethanol in increasing concentrations of 30%, 50%, 75%, 95%, and 100%, 10 min each.

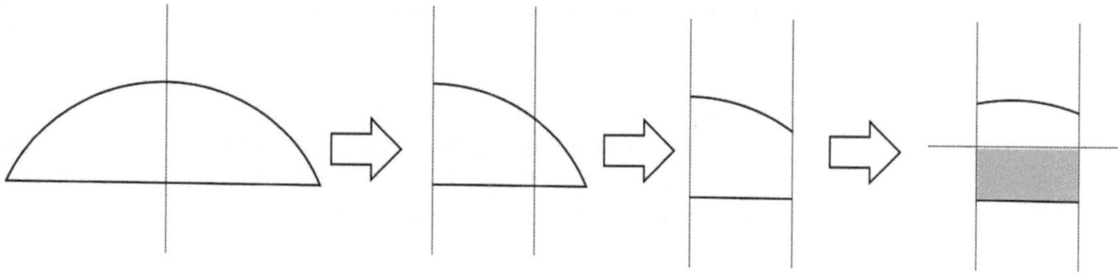

**Fig. 3** Schematic representation of how the solidified agarose droplet should be cut before osmium tetroxide fixation. The appressoria should be located only at the base of the agarose droplet, and thus, only this portion should be carried forward to further experimental steps

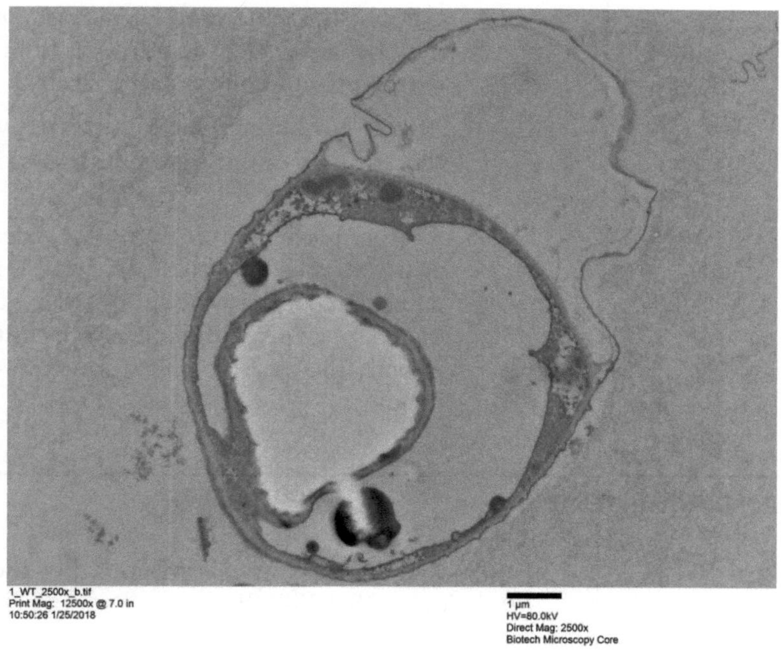


**Fig. 4** TEM image of an appressorium taken after following the described protocol. Note the appressorial pore in the middle of the structure

12. Samples are embedded in blocks using Low Viscosity Embedding Media Spurr's Kit (Electron Microscopy Sciences, Inc., USA).
13. 100 nm thick sections are cut on an Ultramicrotome. Stain sections using uranyl acetate and lead citrate. Samples are now ready for imaging (Fig. 4).

## 4 Notes

1. Cellophane is a flexible surface that can be placed on top of the sample-mounting stub, which reduces the error generated by extra sample handling. However, pre-shape the size of the cellophane to not exceed the limits of the stub.
2. The use of waterproof spray creates a temporary hydrophobic surface on a glass coverslip. Appressoria are able to form in contact with the sprayed layer and are easily removed by scraping with a razor blade.
3. In order to visualize the pore at the base of the appressorium and/or the puncture generated by the penetration peg on the cellophane membrane, 30 min of sonication is the preferred method to detach appressoria from the cellophane surface without forming excessive artifacts.
4. After 1-h fixation in 2.5% glutaraldehyde at room temperature, the assembly can be transferred to a cold room until use for the next steps.
5. Agarose plays the role of a soft holder of the appressoria. The agarose protects the structures from handling-induced damages and keeps the appressoria evenly distributed throughout the agarose, without creating clusters of appressoria, which impede the location and distinction of the specimens during imaging.

## Acknowledgments

Work in the Wilson lab is supported by the National Science Foundation (IOS-1758805).

### References

1. Wilson RA, Talbot NJ (2009) Under pressure: investigating the biology of plant infection by *Magnaporthe oryzae*. Nat Rev Microbiol 7:185–195
2. Yan X, Talbot NJ (2016) Investigating the cell biology of plant infection by the rice blast fungus *Magnaporthe oryzae*. Curr Opin Microbiol 34:147–153
3. Hickey PC, Swift SR, Roca MG, Read ND (2004) Live-cell imaging of filamentous fungi using vital fluorescent dyes and confocal microscopy. Method Microbiol 34:63–87
4. Kankanala P, Czymmek K, Valent B (2007) Roles for rice membrane dynamics and plasmodesmata during biotrophic invasion by the blast fungus. Plant Cell 19:706–724
5. Khang CH, Berruyer R, Giraldo MC, Kankanala P, Park SY, Czymmek K, Kang S, Valent B (2010) Translocation of *Magnaporthe oryzae* effectors into rice cells and their subsequent cell-to-cell movement. Plant Cell 22:1388–1403
6. Giraldo MC, Dagdas YF, Gupta YK, Mentlak TA, Yi M, Martinez-Rocha AL, Saitoh H, Terauchi R, Talbot NJ, Valent B (2013) Two distinct secretion systems facilitate tissue invasion by the rice blast fungus *Magnaporthe oryzae*. Nat Commun 4:1996

7. Marroquin-Guzman M, Hartline D, Wright JD, Elowsky C, Bourret TJ, Wilson RA (2017) The *Magnaporthe oryzae* nitrooxidative stress response suppresses rice innate immunity during blast disease. Nat Microbiol 2:17054
8. Marroquin-Guzman M, Sun G, Wilson RA (2017) Glucose-*ABL1*-TOR signaling modulates cell cycle tuning to control terminal appressorial cell differentiation. PLoS Genet 13:e1006557
9. Sun G, Elowsky C, Li G, Wilson RA (2018) TOR-autophagy branch signaling via Imp1 dictates plant-microbe biotrophic interface longevity. PLoS Genet 14:e1007814
10. Li G, Qi X, Sun G, Rocha RO, Segal LM, Downey KS, Wright JD, Wilson RA (2020) Terminating rice innate immunity induction requires a network of antagonistic and redox-responsive E3 ubiquitin ligases targeting a fungal sirtuin. New Phytol 226:523–540
11. Nayfeh M (2018) Seeing atoms and clusters on surfaces. In: Micro and nano technologies, fundamentals and applications of nano silicon in plasmonics and fullerines. Elsevier, Amsterdam, pp 59–87
12. Corradini MG, DJ MC (2019) Microscopy/Food Applications. In: In Encyclopedia of Analytical Science, 3rd edn. Academic Press, Cambridge, pp 47–56
13. Hand E (2020) Cheap shots. Science 367: 354–358

# Chapter 7

# 4D Widefield Fluorescence Imaging of Appressorium Morphogenesis by *Magnaporthe oryzae*


## Audra Rogers, Nawaraj Dulal, and Martin Egan



## Abstract

Fluorescence microscopy has become a widely used and indispensable tool for the *M. oryzae* research community, providing unique insight into appressorium formation and function. A common practice within the field is to acquire and present images of a number of different conidia, expressing a fluorescent fusion protein of interest, at various stages of infectious development, therein providing a representative "snapshot" of the population at a given point in time. Furthermore, these images typically show only a single focal plane through the specimen (2D) and therefore lack, often valuable, volumetric information. While this approach has its advantages, the continuous imaging of (multiple) single conidia in three dimensions (3D), and over time (4D), can provide additional insight into the spatial and temporal dynamics of fluorescent fusion proteins, and the subcellular structures and compartments they label, in living cells. Here we describe our typical workflow for the 4D live-cell imaging of appressorium morphogenesis in vitro using two-color widefield fluorescence microscopy and briefly outline some important considerations for strain construction, and downstream image processing and visualization.



**Key words** Microscopy, Fluorescence, Imaging, Appressoria, Live-cell imaging, *Magnaporthe oryzae*, Deconvolution


## 1 Introduction

Often, an important first step toward elucidating the function of a protein of interest is to genetically tag it with a fluorescent protein and determine its subcellular localization and dynamic behavior, often relative to other fluorescently labeled proteins, structures, or compartments, in live cells by fluorescence microscopy. Continued advancements in imaging technologies, combined with ever-evolving fluorescent protein tools [1], have allowed for better-than-ever views of the cellular landscape. Fluorescent proteins have been lighting up *Magnaporthe* cells for over two decades [2, 3] and fluorescence microscopy-based imaging continues to provide compelling new insight into appressorium morphogenesis and function. For instance, imaging of the Lifeact-RFP-labeled


Stefan Jacob (ed.), *Magnaporthe oryzae: Methods and Protocols*, Methods in Molecular Biology, vol. 2356, https://doi.org/10.1007/978-1-0716-1613-0_7,

actin cytoskeleton led to the discovery of a toroidal filamentous actin network, templated by a novel higher-order septin ring structure, at the base of the appressorium [4], while careful multicolor time-lapse imaging revealed the unusual spatial uncoupling of cytokinesis from nuclear division during appressorium differentiation [5]. To date, a large and ever-expanding number of *M. oryzae* strains have been generated to image diverse subcellular structures, compartments, and complexes by fluorescence microscopy, including discrete organelles [6–10], cytoskeleton filaments [4, 11], and their associated proteins [5, 12]. Generally speaking, published images of these fluorescently labeled strains are often restricted to two-dimensional (2D) views of a given conidium, acquired at a single focal *Z*-plane, and therefore are missing potentially useful spatial information. Similarly, the now customary in vitro appressorium development time series typically features a different, "representative" conidia imaged at various stages of infectious development, therein providing a "snapshot" of the population at a given timepoint. In contrast, the 3D imaging of fluorescent proteins within (multiple) single conidia, at regular intervals throughout the entire course of appressorium morphogenesis (4D), can provide improved spatial and temporal insight into subcellular protein localization and function (*see* Figs. 1 and 2).

As an organism, *M. oryzae* is particularly well suited to long-term live-cell imaging, displaying considerable tolerance to sustained fluorescence exposure. Furthermore, conidia adhere naturally, and quickly, to cover glass through the release of their own spore tip mucilage, which serves to minimize their movement during multi-dimensional image acquisition. Conidia need only water to germinate and differentiate appressoria and begin to do so within 15 min of their inoculation onto glass surfaces, making *M. oryzae* an unparalleled model system for fluorescence-based live-cell imaging. Here, we describe our typical workflow for the 4D live-cell imaging of appressorium morphogenesis in vitro using multicolor widefield fluorescence microscopy and briefly outline some important considerations for strain construction, and downstream image processing and data visualization.

## 2 Materials

### 2.1 M. oryzae *Culture*

1. Sterile Petri dishes (100 × 15 mm or 60 × 15 mm).
2. Appropriate culture media (e.g., complete media).
3. Laminar flow hood.
4. Tools for subculture (i.e., stainless steel, scissors, scalpels, cork borer, and forceps).

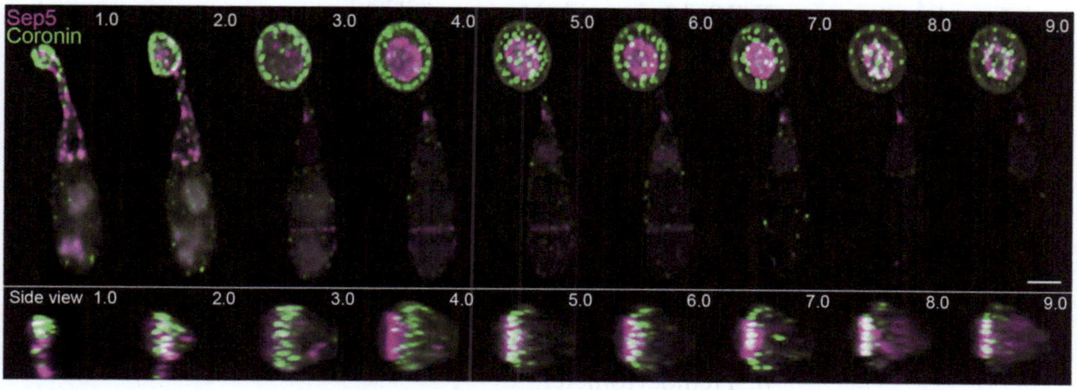


**Fig. 1** Time-resolved 3D widefield fluorescence microscopy-based imaging of the septin cytoskeleton and the actin-binding protein coronin in *M. oryzae* during appressorium development. Images were deconvolved, then displayed as maximum intensity projections of *Z*-stacks. The strain is co-expressing coronin-GFP [12] and Sep5-RFP. Time is shown in hours post-inoculation. Scale = 10 μm

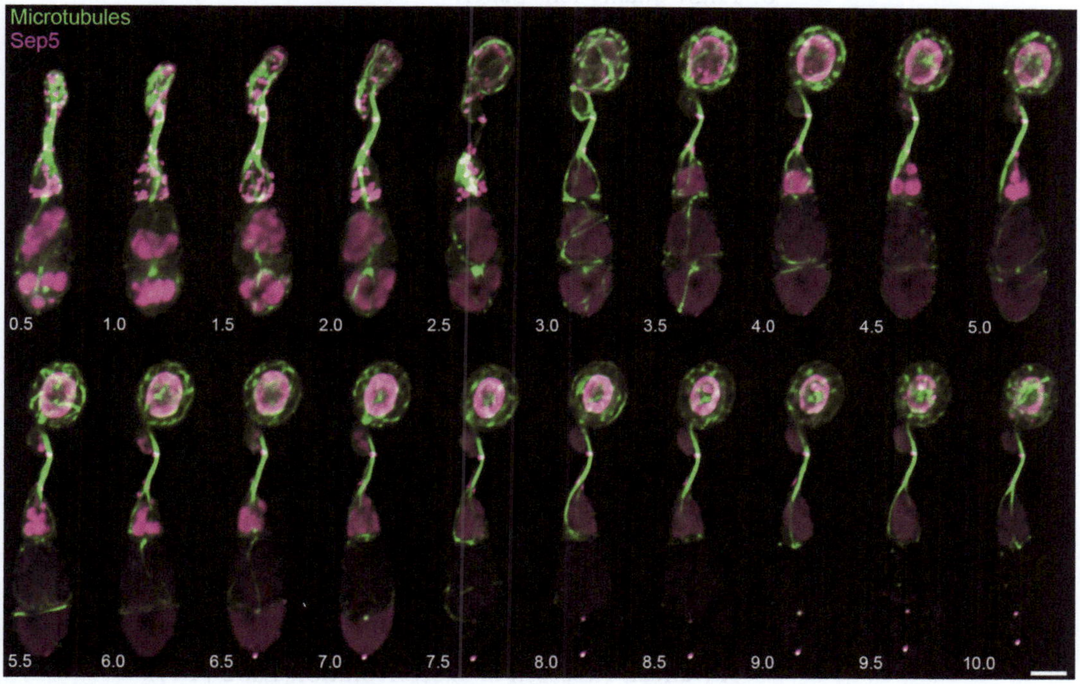


**Fig. 2** Time-resolved 3D widefield fluorescence microscopy-based imaging of the microtubule and septin cytoskeletons in *M. oryzae* during appressorium development. Images were deconvolved, then displayed as maximum intensity projections of *Z*-stacks. The strain is co-expressing tubulin-GFP and Sep5-RFP. Time is shown in hours post-inoculation. Scale = 10 μm

5. Incubator with automatic light settings (e.g., Percival CU-30L2).
6. Desiccated filter paper stocks of relevant *M. oryzae* strains or existing plate culture.

### 2.2 Spore Isolation

1. 10–12-day-old *M. oryzae* plate cultures grown on complete media.
2. Sterilized double distilled deionized water (ddi$H_2O$).
3. Stainless steel cell spreader.
4. Pre-cut and sterilized Miracloth (EMD Millipore).
5. Sterile 2 mL collection tubes (e.g., Eppendorf).
6. Sterile 15 mL or 50 mL conical centrifuge tubes (e.g., Falcon tubes).
7. Microcentrifuge.
8. Hemocytometer.

### 2.3 Live-Cell Imaging Consumables

1. Multi-well borosilicate cell culture chamber slides (e.g., Nunc™ Lab-Tek™ chambers, Thermo Fisher Scientific).
2. Immersion oil (e.g., Nikon Immersion Oil Type F).
3. High-quality lens paper.

### 2.4 Live-Cell Imaging Platform

1. An inverted fluorescence microscope with programmable stage (e.g., Nikon Ti Eclipse Inverted Fluorescence Microscope with encoded motorized XY stage with Piezo Z controlled by NIS-Elements Advanced Research).
2. High-power objective (e.g., Nikon Plan Apo TIRF 100× Oil Immersion Objective Lens NA 1.49 WD 0.13MM).
3. Focus drift correction system (e.g., Nikon Perfect Focus System III or Olympus TruFocus).
4. High-sensitivity camera (e.g., Andor iXon DU-897 Ultra EMCCD).
5. Fluorescence light source (e.g., Lumencor AURA II light engine) and appropriate excitation and emission filters that are well-matched to the spectra of the selected fluorescent proteins.

### 2.5 Image Processing and Data Visualization

1. Access to Fiji [13].
2. Access to deconvolution software (e.g., Nikon NIS-Elements Advanced Research or Deconvolution Lab2 [14]).
3. Access to 3D/4D image visualization tools and software (e.g., Imaris 9.5, Bitplane). For an extended list, *see* [15].

## 3 Methods

All methods should be carried out at room temperature.

### 3.1 General Notes on Construct Design and Strain Construction

1. Before generating DNA constructs and *M. oryzae* strains with new-generation fluorescent proteins (or combinations), confirm that your imaging platform is optimized for their detection and spectral separation. Visit FPbase (https://www.fpbase.org), a community-editable database for fluorescent proteins, for detailed information.
2. Be aware that genetically tagging your protein of interest with a given fluorescent protein may interfere with its normal cellular function [16]. A useful strategy for determining if your fluorescent fusion protein is functional is to test whether it can restore the phenotype of the corresponding gene-deletion strain to that of the wild type when expressed in the mutant background. However, note that successful complementation of the gene-deletion phenotype is no guarantee that localization data is free from tagging artifacts.
3. In the *M. oryzae* field, fluorescent fusion protein constructs are typically introduced ectopically into the genome, often with the endogenous gene left intact. This can mask the effect of non-functional fluorescently tagged proteins and lead to the misinterpretation of data. Wherever possible, fluorescent fusion protein constructs should be integrated at the endogenous loci, in a strain lacking an endogenous copy of the encoding gene, or at "soft landing sites" in the genome [17, 18]. For general guidelines on designing fluorescence microscopy experiments, *see* [16, 19].

### 3.2 M. oryzae Culture

1. Using flame-sterilized tools, cut and place a small (1 $mm^2$) piece of filter paper stock in the center of a CM plate. Alternatively, using a cork borer, subculture a plug of mycelium from the leading edge of an existing culture and transfer to the center of a CM plate.
2. Incubate cultures right-side-up at 25 °C for 10–12 days with a 12:12 light/dark photoperiod to induce sporulation (*see* **Note 1**).
3. To maintain culture vitality, and to prevent the loss of ectopically integrated fluorescent fusion protein constructs, return to frozen filter stocks after no more than two plate-to-plate subcultures.

### 3.3 Isolation of Conidia

1. Perform all work in a laminar flow hood.
2. Pipette 3–5 mL of sterile ddi$H_2O$ onto the surface of the plate culture.
3. Using a flame-sterilized spreader, gently massage the surface of the culture to release the spores into suspension.
4. Pool the spore suspension at the edge of the plate by carefully tilting it.

5. Pipette off the spore suspension and pass it through two layers of sterile Miracloth into a 15 mL conical centrifuge tube (*see* **Note 2**).
6. Wash spores by centrifugation using a fixed angle rotor at 5000 × *g* for 5 min.
7. Completely resuspend pellet in ddi$H_2O$ and repeat centrifugation twice more.
8. Resuspend the pellet in 1 mL of ddi$H_2O$ or more depending on pellet size.
9. Count spores using a hemocytometer to determine the concentration in spores per mL.

### 3.4 Loading Conidia into Imaging Chambers

We perform live-cell imaging experiments in multi-well cell culture chamber slides, allowing us to image multiple strains simultaneously (e.g., a wild-type strain versus a targeted gene-deletion mutant), or determine the effect of various pharmacological agents alongside carrier-only controls.

1. Inoculate ~350 μL of spores at a final concentration of $5 \times 10^4$ spores per mL in ddi$H_2O$ into a single well of an 8-well cell culture chamber slide (*see* **Note 3**).
2. Leave chambers undisturbed for ~30 min to allow the spores to settle and adhere to the bottom of the chamber slide (*see* **Note 4**).
3. Once spores have adhered fully to the chamber surface, remove persistent hyphal fragments by carefully exchanging the ddi$H_2O$ several times using a pipette (*see* **Note 5**).

### 3.5 Automated Image Acquisition

Carefully optimize the precise acquisition settings for a given 4D imaging experiment based upon, among other things, the nature and combination of fluorescent fusion proteins being expressed, the temporal kinetics of the cell biological phenomenon being investigated, as well as the specifications of the imaging platform. These will include optimizing, exposure times, percent light intensities, Z step size, time interval resolution, and imaging duration. As a general rule, we use the shortest exposure times and lowest light intensities that still produce acceptable signal-to-noise levels, therein minimizing the potential for photodamage and toxicity to *M. oryzae* cells. For complete and specific microscopic operational instructions, consult your user manual. Here, we outline only some key steps.

1. Lower the position of the 100× objective by rotating the focus knob.
2. Carefully add a drop of immersion oil onto the clean 100× objective being sure not to touch the lens with the applicator (*see* **Note 6**).

3. Firmly secure the cell culture chamber within the appropriate sample holder (e.g., Universal holder, Prior Scientific) and place onto the stage (*see* **Note 7**).
4. Slowly raise the objective toward the underside of the sample chamber until the oil touches the glass and fills the space between the chamber and the lens.
5. Adjust the focus onto the conidia, which should be static at the bottom of the chamber.
6. Turn on the focus drift correction system.
7. Using the motorized stage and image acquisition interface, program in the spatial coordinates (*X*, *Y*, and *Z*) of multiple fields of view, containing the desired number of conidia (*see* **Note 8**).
8. Set the appropriate time interval in the image acquisition software (e.g., for imaging septin ring formation, we acquire images every 30 min for a total of 12–24 h).
9. Set the appropriate Z step size and define the range (*see* **Note 9**).
10. Set the appropriate exposure times, LED light intensities, and EMCCD gain settings to obtain acceptable signal-to-noise in your images.
11. Start automated image acquisition using the microscope controlling software (*see* **Note 10**).

### 3.6 Image Processing and Data Visualization

1. 4D live-cell imaging results in the generation of fairly large data sets. For example, our multipoint, two-color, 3D time-lapse data sets are typically ~90 GB each. Therefore, consideration should be made toward data storage and transfer infrastructure [20]. Generally, only a subset of imaged conidia are required for downstream analysis, and these can easily be extracted from the main data sets for ease of file transfer and processing.
2. The quality of 3D widefield fluorescence images can be dramatically improved by computational deconvolution, whereby out-of-focus fluorescence, which appears as a blurring, is either subtracted or reassigned to sharpen the image and increase the signal-to-noise ratio [14, 21, 22]. Deconvolution is computationally demanding. We therefore typically deconvolve only a subset of our raw data, and crop fields of view as much as possible to reduce file sizes.
3. While real-time focus correction systems effectively eliminate *Z* drift during long-term time-lapse experiments, a minor amount of *X*–*Y* drift is possible due, for example, to temperature fluctuations or vibrations during the imaging session. This drift between timepoints can be corrected post-acquisition using a number of "image registration" plugins in Fiji [13].

4. We generally perform further image manipulations (e.g., brightness and contrast adjustments) in Fiji. For useful guidelines on the acceptable manipulations of fluorescence microscopy data, *see* [23] (*see* **Note 11**).
5. Be considerate of color-blind individuals when presenting multicolor fluorescence images and do not use green/red combination which will be uninterpretable for those with deuteranopia (red-green color blindness). Green/magenta, blue/yellow, and cyan/red are more acceptable. For further insight, *see* [24].

## 4 Notes

1. Do not use parafilm to seal plates as in our experience this will severely inhibit sporulation.
2. This will remove unwanted hyphal fragments that would otherwise interfere with imaging. For some gene-deletion mutants, these hyphal fragments are more abundant and will persist, but can be largely removed prior to imaging (*see* Subheading 3.3).
3. The stated volume and concentration has been optimized to give an appropriate density of spores at the imaging surface using 8-well Nunc™ Lab-Tek™ chambers and may therefore need to be adjusted for other devices. We limit the number of spores to ~2 per field of view ($81.92 \times 81.92\ \mu m^2$), allowing enough space for germ tube extension and appressorium formation, and the subsequent cropping of discrete infectious units (conidium, germ tube, and appressorium) for automated quantitative analysis and data presentation purposes.
4. This time can be used to turn on the imaging platform and to configure automated image acquisition parameters in the controlling software.
5. Where necessary, apply cutin monomer, 1,16-hexadecanediol, to the conidial suspensions to promote appressorium formation [25].
6. For guidance on the proper use of objective immersion oil, *see* video tutorials at https://nic.med.harvard.edu/oil/.
7. Poorly secured chambers will slide within the sample holder during multipoint imaging resulting in increased *X*–*Y* drift.
8. We typically use the fluorescence signal itself to focus and locate conidia, rather than brightfield, as this allows us to confirm the fluorescent markers and avoid weakly expressing conidia. Select only the conidia that are firmly adhered to the cover glass, and do not move during stage adjustments. Even small movements

of a conidium during *Z*-stack acquisition will result in fluorescence blurring, and loss of resolution, when viewed in three dimensions or as maximum intensity projections.

9. We typically use a *Z* step size of 0.2 μm, spanning a total of 14 μm, which is sufficient to capture the entire depth of the conidium and the subsequent appressorium.
10. Autofluorescence of the appressorium cell wall can be problematic, especially when the expression and relative brightness of a given fluorescent fusion protein is low. When imaging strains expressing a new fluorescent fusion protein, an untransformed wild-type control should be imaged in parallel to help discern autofluorescence from real signal.
11. For many cell biology journals, all nonlinear adjustments to image data (e.g., gamma adjustments) must be disclosed in the figure legends and material and methods sections.

## References


1. Lambert TJ (2019) FPbase: a community-editable fluorescent protein database. Nat Methods 16:277–278
2. Kershaw MJ, Wakley G, Talbot NJ (1998) Complementation of the Mpg1 mutant phenotype in *Magnaporthe grisea* reveals functional relationships between fungal hydrophobins. EMBO J 17:3838–3849
3. Liu Z-M, Kolattukudy PE (1999) Early expression of the calmodulin gene, which precedes appressorium formation in *Magnaporthe grisea*, is inhibited by self-inhibitors and requires surface attachment. J Bacteriol 181:3571–3577
4. Dagdas YF, Yoshino K, Dagdas G, Ryder LS, Bielska E, Steinberg G, Talbot NJ (2012) Septin-mediated plant cell invasion by the rice blast fungus, *Magnaporthe oryzae*. Science 336:1590–1595
5. Saunders DGO, Dagdas YF, Talbot NJ (2010) Spatial uncoupling of mitosis and cytokinesis during appressorium-mediated plant infection by the rice blast fungus *Magnaporthe oryzae*. Plant Cell 22:2417–2428
6. Yang CD, Dang X, Zheng HW, Chen XF, Lin XL, Zhang DM, Abubakar YS, Chen X, Lu G, Wang Z, Li G, Zhou J (2017) Two Rab5 homologs are essential for the development and pathogenicity of the rice blast fungus *Magnaporthe oryzae*. Front Plant Sci 8:620
7. Kershaw MJ, Talbot NJ (2009) Genome-wide functional analysis reveals that infection-associated fungal autophagy is necessary for rice blast disease. Proc Natl Acad Sci U S A 106:15967–15972
8. Patkar RN, Pamplona MR, Gupta AP, Fan Y, Naqvi NI (2012) Mitochondrial β-oxidation regulates organellar integrity and is necessary for conidial germination and invasive growth in *Magnaporthe oryzae*. Mol Microbiol 86:1345–1363
9. Veneault-Fourrey C, Barooah M, Egan M, Wakley G, Talbot NJ (2006) Autophagic fungal cell death is necessary for infection by the rice blast fungus. Science 312:580–583
10. He M, Kershaw MJ, Soanes DM, Xia Y, Talbot NJ (2012) Infection-associated nuclear degeneration in the rice blast fungus *Magnaporthe oryzae* requires non-selective macro-autophagy. PLoS One 7:e33270
11. Park G, Bruno KS, Staiger CJ, Talbot NJ, Xu JR (2004) Independent genetic mechanisms mediate turgor generation and penetration peg formation during plant infection in the rice blast fungus. Mol Microbiol 53:1695–1707
12. Li X, Zhong K, Yin Z, Hu J, Wang W, Li L, Zhang H, Zheng X, Wang P, Zhang Z (2019) The seven transmembrane domain protein MoRgs7 functions in surface perception and undergoes coronin MoCrn1-dependent endocytosis in complex with Gα subunit MoMagA to promote cAMP signaling and appressorium formation in *Magnaporthe oryzae*. PLoS Pathog 15:e1007382
13. Schindelin J, Arganda-Carreras I, Frise E, Kaynig V, Longair M, Pietzsch T, Preibisch S, Rueden C, Saalfeld S, Schmid B, Tinevez J-Y, White DJ, Hartenstein V, Eliceiri K, Tomancak P, Cardona A (2012) Fiji: an

open-source platform for biological-image analysis. Nat Methods 9:676–682

14. Sage D, Donati L, Soulez F, Fortun D, Schmit G, Seitz A, Guiet R, Vonesch C, Unser M (2017) DeconvolutionLab2: an open-source software for deconvolution microscopy. Methods 115:28–41
15. Walter T, Shattuck DW, Baldock R, Bastin ME, Carpenter AE, Duce S, Ellenberg J, Fraser A, Hamilton N, Pieper S, Ragan MA, Schneider JE, Tomancak P, Hériché JK (2010) Visualization of image data from cells to organisms. Nat Methods 7:S26–S41
16. Jost APT, Waters JC (2019) Designing a rigorous microscopy experiment: validating methods and avoiding bias. J Cell Biol 218:1452–1466
17. Yang F, Naqvi NI (2014) Sulfonylurea resistance reconstitution as a novel strategy for ILV2-specific integration in *Magnaporthe oryzae*. Fungal Genet Biol 68:71–76
18. Guo M, Zhu X, Li H, Tan L, Pan Y (2016) Development of a novel strategy for fungal transformation based on a mutant locus conferring carboxin-resistance in *Magnaporthe oryzae*. AMB Express 6:57–59
19. Lee J-Y, Kitaoka M (2018) A beginner's guide to rigor and reproducibility in fluorescence imaging experiments. Mol Biol Cell 29:1519–1525
20. Wallace CT, St Croix CM, Watkins SC (2015) Data management and archiving in a large microscopy-and-imaging, multi-user facility: problems and solutions. Mol Reprod Dev 82:630–634
21. McNally JG, Karpova T, Cooper J, Conchello JA (1999) Three-dimensional imaging by deconvolution microscopy. Methods 19:373–385
22. Sibarita J-B (2005) Deconvolution microscopy. Adv Biochem Eng Biotechnol 95:201–243
23. Johnson J (2012) Not seeing is not believing: improving the visibility of your fluorescence images. Mol Biol Cell 23:754–757
24. Landini G, Perryer DG (2011) More on color blindness. Nat Methods 8:891–892
25. Gilbert RD, Johnson AM, Dean RA (1996) Chemical signals responsible for appressorium formation in the rice blast fungus *Magnaporthe grisea*. Physiol Mol Plant Pathol 48:335–346

# Chapter 8

# Tandem Affinity Purification (TAP) of Low-Abundance Protein Complexes in Filamentous Fungi Demonstrated Using *Magnaporthe oryzae*

**Gang Li and Richard A. Wilson**

## Abstract

Protein–protein interactions underlie cellular structure and function. In recent years, a number of methods have been developed for the identification of protein complexes and component proteins involved in the control of various biological pathways. Tandem affinity purification (TAP) coupled with mass spectrometry (MS) is a powerful method enabling the isolation of high-purity native protein complexes under mild conditions by performing two sequential purification steps using two different epitope tags. In this protocol, we describe a TAP-MS methodology for identifying protein-protein interactions present at very low levels in the fungal cell. Using the 6xHis-3xFLAG double tag, we start the affinity purification process for our protein of interest using high-capacity $Ni^{2+}$ columns. This allows for greatly increased sample input compared to antibody-based first-step purification in conventional TAP protocols and provides a large amount of highly concentrated and preliminarily purified protein complexes to be used in a second purification step involving FLAG immunoprecipitation. The second step greatly facilitates the capture of low-level interacting partners under in vivo conditions. Our TAP-MS method has been proven to secure the characterization of low-abundance protein complexes under physiological conditions with high efficiency, specificity, and economy in the filamentous fungus *Magnaporthe oryzae* and might benefit gene function and proteomics studies in plants and other research fields.

**Key words** Tandem affinity purification, Mass spectrometry, Low-abundance protein complexes, Filamentous fungi, *Magnaporthe oryzae*

## 1 Introduction

Genes encode proteins that often function in the context of protein–protein interactions (PPIs) [1]. PPIs underlie a broad range of biological processes, including physiological and pathological control [2]. Identification of the constituents of protein complexes is crucial to gain insight into the function of the target gene. One of the most widely used methods for uncovering protein–protein interactions is co-immunoprecipitation (co-IP), in which an antibody, directed against an endogenous protein or an epitope-tagged

Stefan Jacob (ed.), *Magnaporthe oryzae: Methods and Protocols*, Methods in Molecular Biology, vol. 2356,
https://doi.org/10.1007/978-1-0716-1613-0_8, 

protein, is used to capture the protein complexes associated with the protein of interest [3]. However, although performing co-IP is straightforward, nonspecific binding of the antibodies to unrelated proteins frequently occurs, especially when co-IP is carried out in batch leading to potentially large numbers of false-positive PPIs [4]. To reduce the amount of nonspecific binding, tandem affinity purification (TAP), performed with two successive purification steps based on two different affinity tags fused to the bait protein (*see* **Note 1**), was first developed in yeast [5] and then used in mammalian cells [6–8], bacteria [9–11], and plants [12–15]. The additional purification step in the TAP method increases the selectivity of the desired protein complexes and greatly reduces the contamination from sticky and highly abundant cellular proteins [16].

Conventional TAP method typically purifies more protein complexes of interest than co-IP and other methods. Sometimes, however, it is still unsatisfactory in isolating very low-abundant protein complexes, even following enrichment processes [5, 7]. For example, the target signal is not visible on Coomassie blue dye or silver-stained SDS-PAGE gel. This is especially a problem for the study of very low-abundant PPIs in filamentous fungi, where protein yields under native cell lysis conditions, in our experience, are much lower than that in plant cells. This is likely due to the thick fungal cell wall that is essential for maintaining structural integrity and is of vital importance for protection against environmental stressors, such as osmotic stress or antifungal substances that disrupt fungal cell wall integrity [17, 18]. One way of tackling very low-abundant PPIs is to increase the initial amounts of tissues used in the cell lysis procedure, but this is limited by the high cost of antibody-coupled beads used in the first-step affinity purification of TAP. To obviate this problem, during a recent study on the regulation of the low-abundant fungal sirtuin protein Sir2 in the filamentous fungus *Magnaporthe oryzae* [19], we experimentally optimized the two affinity purification steps and established an effective TAP methodology for identifying Sir2 interacting partners that control protein stability. This was achieved by fusing a double affinity tag 6xHis-3xFLAG to Sir2. Through TAP purification coupled with mass spectrometry (MS) analysis, we successfully identified a set of E3 ubiquitin protein ligases that are weakly expressed in *M. oryzae* but play essential roles in modulating Sir2 protein abundance under normal and oxidative stress conditions. This was demonstrated to be vital for establishing the biotrophic stage of *M. oryzae* during rice infection. Here, we describe in detail the whole procedure for our optimized TAP purification system, with tips for effective cell lysis, protein complex isolation and enrichment, as well as subsequent data analysis using the filamentous fungus *M. oryzae*. This methodology will facilitate researchers in identifying low-abundant PPI complexes in other fungi and might also be relevant to plant and mammal cells.

## 2 Materials

High-grade chemicals are the preferred choice. General lab equipment such as pipettes, scales, stir sticks, and fume hoods are not mentioned.

### 2.1 Fungal Hyphae Preparation

1. 500 mL graduated cylinder.
2. Narrow-neck 500 mL glass flask bottles.
3. Liquid complete medium (CM). 10 g/L glucose, 2 g/L peptone, 1 g/L yeast extract, 1 g/L casamino acids, 0.1% (v/v) trace elements, 0.1% (v/v) vitamin supplement, 6 g/L $NaNO_3$, 0.5 g/L KCl, 0.5 g/L $MgSO_4$, 1.5 g/L $KH_2PO_4$; adjust to PH 6.5 with 10 M NaOH, and then sterilize by autoclaving at 121 °C for 30 min [20].
4. Trace elements (working concentration for each component in CM: g/mL).

   $ZnSO_4 \times 7H_2O$ (0.0022%), $H_3BO_3$ (0.0011%), $MnCl_2 \times 4H_2O$ (0.0005%), $FeSO_4 \times 7H_2O$ (0.0005%), $CoCl_2 \times 6HOH$ (0.00017%), $CuSO_4 \times 5H_2O$ (0.00016%), $Na_2MoO_4 \times 2H_2O$ (0.00015%), $Na_4EDTA$ (0.005%).
5. Vitamin supplement (working concentration for each component in CM: g/mL).

   Biotin (0.0001%), pyridoxine (0.0001%), thiamine (0.0001%), riboflavin (0.0001%), PABA (*p*-aminobenzoic acid) (0.0001%), nicotinic acid (0.0001%).
6. Orbital incubator shaker.
7. Miracloth.
8. Sterile distilled water.
9. Tissue paper.
10. Liquid nitrogen.

### 2.2 Total Protein Preparation for TAP

1. Cold room (4 °C).
2. Mortar and pestle.
3. 50 mL conical centrifuge tubes.
4. 15 mL conical centrifuge tubes.
5. PE buffer. 50 mM Tris–HCl (pH 8.0), 300 mM NaCl, 10% glycerol, 10 mM imidazole, 1× protease inhibitor cocktail, 1× phosphatase inhibitor cocktail, 50 μM MG132, and 1 mM phenylmethylsulfonyl fluoride (PMSF) in water (*see* **Note 2**).
6. Vortex mixer.
7. High-speed centrifuge tubes.
8. Ultracentrifuge (e.g., Beckman Coulter).

9. 1.5 mL low-retention microcentrifuge tubes (e.g., Protein LoBind conical tubes).
10. Protein quantification kit (e.g., Bradford Protein Assay Reagent).

### 2.3 $Ni^{2+}$ Affinity Chromatography

1. $Ni^{2+}$ affinity column (e.g., HisTrap™ High Performance, 1 mL, GE Healthcare).
2. HW buffer. 50 mM Tris–HCl (pH 8.0), 300 mM NaCl, 10% glycerol, 10 mM imidazole in water (*see* **Note 3**).
3. HE buffer. 50 mM Tris–HCl (pH 8.0), 300 mM NaCl, 10% glycerol, 500 mM imidazole, 1× protease inhibitor cocktail, 1× phosphatase inhibitor cocktail, 50 μM MG132, in water (*see* **Notes 2** and **4**).
4. 1.5 mL low-retention microcentrifuge tubes (e.g., Protein LoBind conical tubes).

### 2.4 FLAG Immunoprecipitation

1. 1.5 mL low-retention microcentrifuge tubes (e.g., Protein LoBind conical tubes).
2. ANTI-FLAG M2 Affinity Agarose Gel.
3. Benchtop centrifuge with temperature control.
4. Tube rotator.
5. 1× TBS (pH 7.5). 50 mM Tris-HCl (pH 7.5), 150 mM NaCl in water.
6. 0.1 M glycine–HCl (pH 3.5). 0.1 M glycine–HCl in water; adjust to pH 3.5 with HCl.
7. FW buffer. 50 mM Tris–HCl (pH 8.0), 150 mM NaCl, 10% glycerol, 1 mM EDTA, 1× protease inhibitor cocktail, 1× phosphatase inhibitor cocktail in water.
8. 3× FLAG peptide solution. 300 ng/μL 3× FLAG peptide, 50 mM Tris–HCl (pH 7.5), 150 mM NaCl in water.
9. Neutralization buffer. 0.5 M Tris–HCl (pH 7.4), 1.5 M NaCl in water.
10. 2× LDS sample buffer. 20% glycerol, 2% lithium dodecyl sulfate (LDS), 0.4 M triethanolamine–HCl (pH 7.6), 2% Ficoll 400, 0.0125% phenol red, 0.0125% Coomassie G250, 1 mM EDTA in water.
11. Heated water bath at 95 °C.

### 2.5 Data Analysis

1. Search engine (e.g., Mascot).
2. Protein database for *M. oryzae* (e.g., downloaded from the Broad Institute (https://www.broadinstitute.org/genomics) or the UniProt (www.uniprot.org)).
3. Protein database for common contaminants (e.g., the cRAP).

4. A custom protein database containing the user-supplied fusion protein sequence.
5. Proteome software (e.g., Scaffold).
6. Statistical analysis tools (e.g., Microsoft Excel, R, or Python).

## 3 Methods

PPI is a function-driven molecular event. To perform specific biological functions, proteins interact with their partners dynamically and transiently [21], so the temporal nature of protein interactions, the decisive factor of the components of protein complexes, should be considered prior to PPI investigation. For example, in our recent paper [19], a time course expression study under oxidative stress conditions was first carried out, and then, based on those results, the time point at 40 min after $H_2O_2$ treatment was chosen as the optimal condition for *SIR2* complex isolation. In addition, the spatial expression patterns of the gene of interest (GOI) should be determined before TAP, which is necessary for tissue-specific expressed genes. In this respect, we used vegetative hyphae for preparing total proteins for TAP because we found the *SIR2* gene was expressed in the vegetative hyphae under both normal and oxidative stress conditions [19]. In addition to gene transcriptional analysis, the solubility of the protein needs to be considered for optimizing TAP, especially for membrane-associated proteins, for which the concentration of non-denaturing detergents should be increased to break protein–lipid interactions. Furthermore, it is better to incorporate a control protein, e.g., green fluorescent protein (GFP), in the whole process of TAP, in order to differentiate background proteins during data analysis.

### 3.1 Fungal Hyphae Preparation

1. Inoculate the fungal strains expressing the GOI (gene of interest) fused to *6xHis-3xFLAG*, and a control stain expressing $GFP^{6H3F}$, in 350 mL liquid CM, and grow at 25 °C for 48 h with continuous shaking at 140 rpm (*see* **Note 5**).
2. Harvest the mycelium with Miracloth, wash excessively with sterile distilled water, pat dry with tissue paper, flash freeze in liquid nitrogen, and store at −80 °C until use.

### 3.2 Total Protein Preparation for TAP

Perform all steps from now on in a cold room with the temperature set to 4 °C unless stated otherwise.

1. Grind the frozen fungal mycelium to fine powder in liquid nitrogen using a prechilled mortar and pestle.
2. For each strain, carefully transfer 12 g of finely ground mycelium to a prechilled 50 mL conical centrifuge tube followed by

adding precold 20 mL of PE buffer. Mix gently with a sterile stirring rod followed by a short vortexing which can help to homogenize the mycelium in the PE buffer (*see* **Note 6**).

3. Transfer the mixture to a high-speed centrifuge tube, and ultracentrifuge at >20,000 × *g* at 4 °C for 30 min.
4. After removal of insoluble cell debris by centrifugation, transfer the supernatant (crude extract) to a precold 50 mL conical centrifuge tube. Save 100 μL as the loading control (L) for $Ni^{2+}$ affinity chromatography.
5. Measure protein concentration using Bradford Protein Assay. For the test strain and the control strain, adjust the protein concentration to a comparable level with PE buffer (*see* **Note 7**).

### 3.3 $Ni^{2+}$ Affinity Chromatography

1. Wash the 1 mL HisTrap HP column first with 5 mL of sterile distilled water and then equilibrate with 10 mL of HW buffer (*see* **Note 8**).
2. Load the crude extract through the equilibrated column into another 50 mL conical tube on ice. Save the flow-through for quality controls (His-FT).
3. Wash the column with 20 mL of PE buffer. Save the wash for quality controls (His-W).
4. Sequentially elute bound proteins off the column with 5 mL of HE buffer into ten 1.5 mL low-retention tubes (E1–E10) with 500 μL per fraction. Save 10 μL of each fraction for quality controls. Place the samples E1–E10 on ice until use.
5. Determine in an immunoblot analysis of E1–E10 using anti-FLAG antibody the fractions in which the His-FLAG-tagged proteins are enriched (*see* **Note 9**).
6. Pool two fractions containing most of the enriched His-FLAG-tagged proteins as FLAG-IP loading and proceed to FLAG immunoprecipitation (*see* **Note 10**). Save 10 μL of FLAG-IP loading for quality controls.

### 3.4 FLAG Immunoprecipitation

1. Remove 50 μL ANTI-FLAG M2 Affinity Agarose Gel (bead slurry) to a 1.5 mL low-retention tube (*see* **Note 11**).
2. Wash the resin twice with 1 mL of 1× TBS (pH 7.5) followed by 0.5 mL of 0.1 M glycine–HCl (pH 3.5) once to remove any trace of amounts of unbound antibody from the resin suspension (*see* **Note 12**). Next, wash the resin three times with 1 mL of 1× TBS (pH 7.5) (*see* **Note 13**). Between the washing steps, spin down the resin at 3000 × *g* for 30 s at 4 °C. Carefully remove the supernatant with a pipette after each centrifugation.

3. Add 1 mL of HE buffer to the resin for equilibration and then pellet the beads by spinning at 3000 × *g* for 30 s at 4 °C. Carefully remove the supernatant with a pipette.
4. Add the combined two fractions from Subheading 3.3, **step 6** to the washed resin. Agitate all samples and controls gently at 4 °C for 2 h by end-over-end rotation on a tube rotator.
5. At the end of incubation, pellet the resin at 3000 × *g* for 30 s at 4 °C. Carefully remove the flow-through to another 1.5 mL low-retention tube without touching the resin. Save 4 μL of flow-through for quality controls (FLAG-FT).
6. Wash the resin four times in 500 μL of FW buffer and then wash four times in 500 μL of 1× TBS buffer (pH 7.5). Save 4 μL of the first (W1) and the second (W2) wash for quality control. For each wash, gently resuspend the resin with wash buffer followed by gentle rotation for 5 min at 4 °C on a roller shaker.
7. Centrifuge the resin at 3000 × *g* for 30 s at 4 °C. Carefully remove the supernatant using a pipette without taking any resin.
8. Elute immunoprecipitated proteins of FLAG resin with either 3× FLAG peptide solution, with 0.1 M glycine–HCl (pH 3.5), or with 2× LDS sample buffer (*see* **Note 14**). Save 1 μL of each eluate for quality controls (FLAG-E). Store the eluate at 4 °C for immediate use or −20 °C for long-term storage.
   (a) *Elution with 3× FLAG peptide solution*: add 100 μL of 3× FLAG peptide solution (300 ng/μL) to the resin followed by gentle shaking for 30 min at 4 °C, and then centrifuge the resin at 8200 × *g* for 30 s at 4 °C. Carefully transfer the supernatant to a fresh 0.5 mL low-retention tube labeled with FLAG-$E_{pep}$. Save 4 μL of the sample for quality controls.
   (b) *Elution with 0.1 M glycine (pH 3.5)*: add 100 μL of 0.1 M glycine (pH 3.5) to the resin followed by gentle shaking for 5 min at room temperature, and then centrifuge the resin at 8200 × *g* for 30 s. Carefully transfer the supernatant to a fresh 0.5 mL low-retention tube containing 10 μL of neutralization buffer. Label the tube with FLAG-$E_{acid}$ and save 4 μL for quality controls.
   (c) *Elution with 2× LDS sample buffer*: add 50 μL of 2× LDS sample buffer (*see* **Note 15**) to the resin followed by incubating in the water bath at 95 °C for 3 min, and then centrifuge the resin at 8200 × *g* for 30 s. Carefully transfer the supernatant to a fresh 0.5 mL low-retention tube labeled with FLAG-$E_{LDS}$. Save 4 μL of the sample for quality controls.

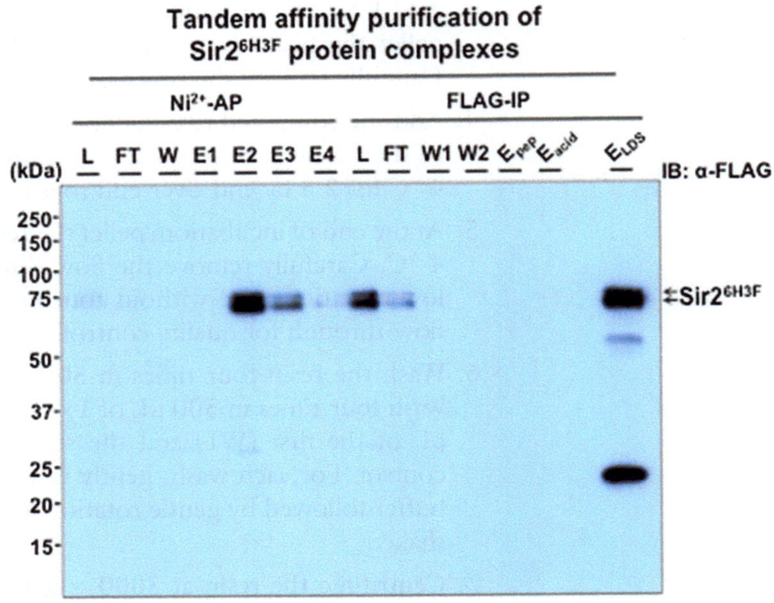


**Fig. 1** Quality evaluation of tandem affinity purification procedures for isolating Sir2-associated protein complexes. Vegetative hyphae of the *Δsir2* complementation strain expressing *SIR2*$^{6H3F}$ were cultured in liquid CM at 25 °C for 48 h and then subjected to oxidative stress treatment with 12 mM $H_2O_2$ for 40 min. Following harvesting of the hyphae, total proteins were isolated with native lysis buffer containing the protease inhibitor cocktail, phosphatase inhibitor cocktail, and the proteasome inhibitor MG-132. After protein preparation, tandem affinity purification was carried out using $Ni^{2+}$ affinity purification followed by FLAG-immunoprecipitation. Note that, based on our experience of the fraction distribution patterns of Sir2$^{6H3F}$ in the eluates of $Ni^{2+}$-AP characterized by the pilot experiment, elution fractions E2 and E3 were directly chosen, combined, and used in the subsequent FLAG-IP directly after the $Ni^{2+}$-AP. The image shown here is the result of immunoblotting analysis of the purification quality of the Sir2$^{6H3F}$ protein complexes throughout our TAP workflow. L, FT, W, and E indicate loading, flow-through, wash, and eluate, respectively; Epep, Eacid, and $E_{LDS}$ denote elution with 3× FLAG peptide solution (300 ng/μL in 1×TBS), acidic condition (0.1 M glycine, pH 3.5), and 2× LDS sample buffer, respectively. Please note that Sir2$^{6H3F}$ protein complexes can only be eluted off FLAG resin by 2× LDS sample buffer but not the 3× FLAG peptide solution or the 0.1 M glycine, pH 3.5. *IB* immunoblotting. (This image is an overexposed version of the one used in Fig. S5 of our recent paper [19])

9. After the FLAG immunoprecipitation, evaluate the quality of the TAP and the efficiency of the final elution of the FLAG-IP by analyzing the saved small aliquots of each fraction using immunoblotting analysis (an example is given in Fig. 1). If the quality and amount is satisfactory, analyze the majority of the FLAG-E fraction by Coomassie blue staining, silver staining, or mass spectrometry (*see* **Note 16**). In this case, use the control protein (e.g., GFP$^{6H3F}$) for differentiating

nonspecifically bound background proteins and thus facilitate the identification of the interaction partners of the tagged $GOI^{6H3F}$.

### 3.5 Data Analysis

1. After the MS data are collected, analyze all MS/MS samples using Mascot or other search engines. These softwares use mass spectrometry data to perform database search for identification of proteins from peptide sequence databases. For identification of *M. oryzae* proteins, search against the protein database MGx in the Broad Institute (https://www.broadinstitute.org/genomics) or UniProt (www.uniprot.org), along with a database for common contaminants, e.g., the cRAP, and a custom database containing the user-supplied fusion protein sequence . For any search engine used, related search parameters, e.g., the digestion enzyme(s) used to generate the peptides prior to MS analysis, the fragment ion mass tolerance, and the parent ion tolerance, should be set up so as to filter out the low-scoring matches and remove the redundancy, all of which can greatly affect the final output.
2. Following database search, use proteome software, e.g., Scaffold, to validate MS/MS-based peptide and protein identifications. In the case of the Scaffold, peptide identifications are accepted if they could be established at greater than 95% probability by the PeptideProphet algorithm [22] with Scaffold delta-mass correction. Protein identifications are accepted if they could be established at greater than 95% probability and contain at least 5–10 identified peptides (*see* **Note 17**). Protein probabilities are assigned by the ProteinProphet algorithm [23]. Proteins that contain similar peptides and cannot be differentiated based on MS/MS analysis alone are grouped to satisfy the principles of parsimony. Proteins sharing significant peptide evidence are grouped into clusters.

   Proteins identified through TAP may be involved in direct or indirect protein-protein interactions; if the direct interaction needs to be determined, other experimental strategies, e.g., yeast two-hybrid analysis, should be further employed. Moreover, as MS analysis is based on peptide identification, confirm authentic interactions using the in vivo pull-down assay between the bait and the prey proteins and antibody-based immunoblotting analysis. In addition, it is worth noting that the extra purification step in TAP increases the manipulation time period which may result in the loss of transient interaction partners; thus, interactive review of the TAP and co-IP data is suggested for a comprehensive understanding of the PPI networks of interest [19].

## 4 Notes

1. Two different tags used for TAP are usually fused at one side (N- or C-terminus) of the bait protein. The fusion should be tested first by complementation assays in a respective mutant background, and only the strain with the functional fusion can be used for TAP.
2. Protease inhibitor cocktail, phosphatase inhibitor cocktail, and MG132 should be EDTA- or EGTA-free as chelating agents will disturb the $Ni^{2+}$ affinity chromatography. PMSF should be freshly prepared and added just before use.
3. The NaCl concentration in the HW buffer can be set to 150–300 mM depending on the stability of the target protein complexes.
4. The NaCl concentration in the HE buffer can be set to 150–300 mM depending on the stability of the target protein complexes.
5. In our case, we fused the tandem affinity tags 6xHis-3xFLAG at the C-terminus of our bait protein Sir2 and used it for TAP as we verified $Sir2^{6H3F}$ is functional in complementing the defects of the pathogenicity of Δ*sir2* knockout mutant strains.
6. Avoid sonication during homogenization of the mycelium because the heat gradient from the probe may damage proteins and disrupt PPIs under non-crosslinking conditions.
7. Adjusting the protein concentration of the test strain and control strain to comparable levels will help differentiate background proteins identified in TAP.
8. Use a syringe or a pump for protein purification using a 1 mL HisTrap HP column. Suggested flow rates for using a 1 mL HisTrap HP column throughout the whole process of $Ni^{2+}$ affinity chromatography are 0.5–1 mL/min.
9. To save time for this step, perform a pilot experiment to determine the tagged protein distribution in elution fractions because the fraction distribution of a specific protein is repeatable if the elution volume is close to 500 μL. Target fractions expected to contain the enriched tagged proteins can be chosen directly for FLAG-IP. Confirm the protein distribution later by immunoblotting with the saved 10 μL aliquot of each fraction.
10. According to our experience, the 500 mM imidazole in the HE buffer does not interfere with the FLAG-IP of interacting proteins significantly, so imidazole does not have to be removed here.
11. When handling agarose beads, it is better to cut the end of the pipette tips before taking the beads to ensure the uptake of the correct volume of the beads.

12. Do not leave the beads in glycine–HCl for more than 20 min. Remove the glycine–HCl as soon as possible after washing.
13. To carefully remove the supernatant after each centrifugation without losing any resin, narrow-end pipette tips are suggested, which can be made using forceps to pinch the opening of the pipette tip until it is partially closed.
14. Considering the difference in protein characteristics, it is better to try eluting the tagged proteins with various elution conditions as sometimes 3× FLAG peptide or 0.1 M glycine may fail in elution of target protein off the agarose resin (see an example given in Fig. 1).
15. To minimize the denaturation and elution of the immobilized M2 antibody, do not include reducing agents, including 2-mercaptoethanol or DTT, in the LDS sample buffer.
16. If the purified proteins are eluted using anionic detergents, e.g., SDS or LDS, remove the detergents prior to LC-MS/MS analysis. This can be achieved by running an SDS-PAGE followed by in-gel digestion, or by TCA precipitation, etc. If the in-solution digestion is preferred before LC-MS/MS, use the acid labile surfactants, e.g., RapiGest SF, PPS Silent Surfactant, or ProteaseMAX for protein elution from FLAG resin. These surfactants are compatible with the most commonly used proteolytic enzyme trypsin and can be easily removed through hydrolysis in acidic conditions before LC-MS/MS.
17. For regular one-step affinity purification, e.g., co-IP, the protein identifications can be accepted if they contain at least two identified peptides. However, for TAP, it is better to drive up the standards, i.e., the identified proteins should contain at least 5–10 identified peptides, as the constituents of the protein complex are enriched by an additional purification step in TAP, though the expression level of the proteins should also be considered.

## Acknowledgments

This work was supported by the National Science Foundation (IOS-1758805).

### References

1. Jacq B (2001) Protein function from the perspective of molecular interactions and genetic networks. Brief Bioinform 2:38–50
2. Koh GC, Porras P, Aranda B, Hermjakob H, Orchard SE (2012) Analyzing protein-protein interaction networks. J Proteome Res 11:2014–2031
3. Lee C (2007) Coimmunoprecipitation assay. Methods Mol Biol 362:401–406
4. Solstad T, Bull VH, Breivik L, Fladmark KE (2008) Identification of protein interaction partners by immunoprecipitation: possible pitfalls and false positives. E&C Hepatology 4:21–27

5. Rigaut G, Shevchenko A, Rutz B, Wilm M, Mann M, Seraphin B (1999) A generic protein purification method for protein complex characterization and proteome exploration. Nat Biotechnol 17:1030–1032
6. Cox DM, Du M, Guo X, Siu KW, McDermott JC (2002) Tandem affinity purification of protein complexes from mammalian cells. BioTechniques 33:267–268, 270
7. Bürckstümmer T, Bennett KL, Preradovic A, Schütze G, Hantschel O, Superti-Furga G, Bauch A (2006) An efficient tandem affinity purification procedure for interaction proteomics in mammalian cells. Nat Methods 3:1013–1019
8. Tsai A, Carstens RP (2006) An optimized protocol for protein purification in cultured mammalian cells using a tandem affinity purification approach. Nat Protoc 1:2820–2827
9. Gully D, Moinier D, Loiseau L, Bouveret E (2003) New partners of acyl carrier protein detected in *Escherichia coli* by tandem affinity purification. FEBS Lett 548:90–96
10. Fodor BD, Kovács AT, Csáki R, Hunyadi-Gulyás E, Klement E, Maróti G, Mészáros LS, Medzihradszky KF, Rákhely G, Kovács KL (2004) Modular broad-host-range expression vectors for single-protein and protein complex purification. Appl Environ Microbiol 70:712–721
11. Kumar JK, Tabor S, Richardson CC (2004) Proteomic analysis of thioredoxin-targeted proteins in *Escherichia coli*. Proc Natl Acad Sci U S A 101:3759–3764
12. Rivas S, Romeis T, Jones JD (2002) The Cf-9 disease resistance protein is present in an approximately 420-kilodalton heteromultimeric membrane-associated complex at one molecule per complex. Plant Cell 14:689–702
13. Rohila JS, Chen M, Cerny R, Fromm ME (2004) Improved tandem affinity purification tag and methods for isolation of protein heterocomplexes from plants. Plant J 38:172–181
14. Rubio V, Shen Y, Saijo Y, Liu Y, Gusmaroli G, Dinesh-Kumar SP, Deng XW (2005) Plant J 41:767–778
15. Ham BK, Li G, Kang BH, Zeng F, Lucas WJ (2012) Overexpression of *Arabidopsis* plasmodesmata germin-like proteins disrupts root growth and development. Plant Cell 24:3630–3648
16. Puig O, Caspary F, Rigaut G, Rutz B, Bouveret E, Bragado-Nilsson E, Wilm M, Séraphin B (2001) The tandem affinity purification (TAP) method: a general procedure of protein complex purification. Methods 24:218–229
17. García R, Bermejo C, Grau C, Pérez R, Rodríguez-Peña JM, Francois J, Nombela C, Arroyo J (2004) The global transcriptional response to transient Cell Wall damage in *Saccharomyces cerevisiae* and its regulation by the cell integrity signaling pathway. J Biol Chem 279:15183–15195
18. Meyer V, Damveld RA, Arentshorst M, Stahl U, van den Hondel CA, Ram AF (2007) Survival in the presence of antifungals: genome wide expression profiling of *Aspergillus niger* in response to sublethal concentrations of caspofungin and fenpropimorph. J Biol Chem 282:32935–32948
19. Li G, Qi X, Sun G, Rocha RO, Segal LM, Downey KS, Wright JD, Wilson RA (2020) Terminating rice innate immunity induction requires a network of antagonistic and redox-responsive E3 ubiquitin ligases targeting a fungal sirtuin. New Phytol 226(2):523–540
20. Li G, Marroquin-Guzman M, Wilson RA (2015) Chromatin immunoprecipitation (ChIP) assay for detecting direct and indirect protein–DNA interactions in *Magnaporthe oryzae*. Bio-protocol 5:21
21. Nooren I, Thornton JM (2003) Diversity of protein–protein interactions. EMBO J 22:3486–3492
22. Keller A, Nesvizhskii AI, Kolker E, Aebersold R (2002) Empirical statistical model to estimate the accuracy of peptide identifications made by MS/MS and database search. Anal Chem 74:5383–5392
23. Nesvizhskii AI, Keller A, Kolker E, Aebersold R (2003) A statistical model for identifying proteins by tandem mass spectrometry. Anal Chem 75:4646–4658

# Chapter 9

# Quantitative Proteome and Phosphoproteome Profiling in *Magnaporthe oryzae*

**Thomas Michna and Stefan Tenzer**


## Abstract

The quality and consistency in every sample preparation procedure is crucial for any scientific output. Therefore, it is of utmost importance to have easy, economic, and robust sample preparation protocols. Here, we describe a simple and robust bottom-up proteomic sample preparation strategy for identification and label-free quantification (LFQ) of proteins and phosphoproteins. The presented workflow is designed for large-scale application and involves easy scalable and well-known robust sample preparation techniques, such as cell lysis with SDS buffer under heat, protein precipitation using methanol/chloroform, tryptic digest, and commercially available $TiO_2$ phosphopeptide enrichment kits. Over a sample set of 48 samples of only 200 mg fungal mycelium each, we quantified a median of 2937 proteins after processing in the IsoQuant software. The median peptide count was 10 peptides per protein leading to a median 65% sequence coverage. In addition, we identified a median of 3324 phosphopeptides (corresponding to 998 phosphoproteins) with 4874 phosphosites per sample. Over all samples, we achieved a median phosphopeptide enrichment efficiency of 77%. The distribution of serine/threonine/tyrosine (S/T/Y) phosphosites was 78.1%/21.2%/0.6%.


**Key words** Proteomics, Label-free quantification, *Magnaporthe oryzae*, $TiO_2$, Phosphoproteomics, Phosphopeptide enrichment, Sample preparation

## 1 Introduction

For a comprehensive understanding of complex biological processes, it is necessary to link information of multiple levels, such as transcriptome, proteome, and metabolome [1]. Until 15 years ago, the science of proteomics was very tedious and rather insensitive: For protein identification it was necessary to prepare huge two-dimensional gels (2D-GE) that were difficult to handle, time consuming, and low resolution and were difficult to reproduce. Identification and quantification of single proteins could be done using more or less specific antibodies or other techniques that came either with safety issues or error prone and thus far from robust [2]. The formerly relatively expensive mass-spectrometric (MS)-based identification suffered from high instrument cycle times and

Stefan Jacob (ed.), *Magnaporthe oryzae: Methods and Protocols*, Methods in Molecular Biology, vol. 2356,
https://doi.org/10.1007/978-1-0716-1613-0_9, 

was only operable by highly specialized staff. Nevertheless, during the last years the instrument prices went down and the operability was simplified, making MS-based proteomics the method of choice for global, comprehensive protein analysis. Current generation instruments reproducibly quantify thousands of proteins with high sensitivity, throughput, and robustness, rendering them superior to classical 2D gel approaches in most aspects [3].

Since protein analysis by MS became popular, the basic principle of the strategy remained the same: Proteins undergo digestion by proteases like trypsin and Lys-C yielding smaller peptides. Following separation by reversed-phase liquid chromatography (RP-LC) and electrospray ionization (ESI), the peptide mixtures are analyzed online by the mass spectrometer. In nanoscale ultra-high-performance liquid chromatography (nanoUPLC) systems, peptides are separated by gradients (typically between 30 and 180 min length) and elute over a short time of 5–30 s into the MS. In data-dependent acquisition (DDA), the mass over charge ratios (m/z) and intensities of the eluting peptides are measured first. Subsequently, the most intense signals are selected for fragmentation in the collision cell and fragmented into smaller peptide fragments and amino acids. Bioinformatic tools allow to reconstruct the underlying peptide sequences from the obtained intact and fragment m/z values. Finally, proteins can be identified by mapping the peptide sequence to entries from protein databases. In label-free quantification, intensity information or spectral counts of the peptides are used for peptide and protein quantification [4, 5].

The analysis of phosphorylated peptides is essential for the elucidation of signal transduction pathways, but remains challenging. Due to their low abundance in the peptidome, phosphopeptides will be difficult to detect in the presence of the signals that derive from non-phosphorylated species. Additionally, during electrospray ionization, phosphopeptides ionize less efficiently, which causes low efficiency in simultaneous identification of non- and phosphorylated peptides. Last but not least, the neutral loss of the phospho-group during peptide fragmentation process makes the correct identification of the phosphorylation site difficult [6, 7].

Therefore, for the comprehensive analysis of the phosphoproteome, multiple additional sample preparation steps for phosphopeptide enrichment are necessary. These include immunoprecipitation (IP), metal oxide/immobilized metal ion affinity chromatography (MOAC/IMAC), and fractionation strategies such as high-pH reversed-phase chromatography (high pH RP), strong cation exchange (SCX), or electrostatic repulsion hydrophilic interaction chromatography (ERLIC) [8, 9]. The workflow presented here relies on the commercially available phosphopeptide enrichment by MOAC using titanium dioxide ($TiO_2$) as retaining agent. In short, $TiO_2$ material is filled into standard

200 μL pipetting tips or spin columns. During centrifugation of tryptic digest from the samples through these tips/columns under acidic conditions, the phosphopeptides are retained by the $TiO_2$ material. After washing steps, the phosphopeptides are eluted with basic eluent for further desalting and MS analysis [10].

Few publications addressing the proteomic profiling of *Magnaporthe oryzae* and the consecutive proteomic responses in the rice plant are available. While some rely on traditional 2D-GE approaches [11, 12], others make use of advanced multiple sequentially applied phosphopeptide enrichment strategies [13, 14]. The procedure described here is a tradeoff between accuracy, speed, sample consumption, and simplicity. We utilize a robust in-solution tryptic digestion without the use of cost-intensive inhibitory reagents, yielding a robust and easy scalable workflow for label-free quantitative proteomics and phosphoproteomics.

## 2 Materials

All reagents should be prepared freshly, especially those containing volatile components such as trifluoroacetic acid (TFA) or acetonitrile (ACN). Also, light-sensitive and oxidizing agents such as dithiothreitol (DTT) or iodoacetamide (IAA) should be dissolved right before use, and IAA should be stored with light protection between weighing and solvation. Solvents containing ACN should be degassed using ultrasound treatment. All chemicals should have the highest grade possible; this workflow was tested with HPLC grade reagents. Standard lab equipment such as pipettes and lab scales are not mentioned.

### 2.1 Lysis of Mycelium and Alkylation

1. Lyophilizing equipment.
2. Mortar and pestle.
3. Lysis buffer. 100 mM Tris–HCl (pH 8.5), 5% SDS, 5 mM DTT in water.
4. 50 mL Protein LoBind conical tubes.
5. Alkylation reagent. 200 mM iodoacetamide (IAA), 50 mM AMBIC in water.
6. Heated water bath at 95 °C.
7. Benchtop centrifuge with temperature control (e.g., Heraeus Multifuge 3L-R).

### 2.2 Protein Precipitation, Resuspension, and Quantification

1. Methanol.
2. Chloroform.
3. Water.
4. 50 mL Protein LoBind conical tubes.

5. Resuspension buffer. 7 M urea, 2 M thiourea, 100 mM ammonium bicarbonate (AMBIC) in water.
6. Protein quantification kit (e.g., Pierce 660 nm Protein Assay Reagent).
7. Detection equipment for the protein quantification kit, usually a regular UV/Vis-spectrometer.

### 2.3 In-Solution Tryptic Digestion

1. Dilution buffer. 50 mM AMBIC in water.
2. Digestion reagent. 5 μg/μL trypsin (e.g., TPCK treated), 50 mM acetic acid in water. Aliquots can be stored for 3 months at −20 °C.
3. Trifluoroacetic acid (TFA).

### 2.4 Desalting I

1. Methanol.
2. Elution solvent. 50% acetonitrile (ACN), 0.1% TFA in water.
3. Wash solvent. 0.1% TFA in water.
4. Suitable C18 SPE material (e.g., Sep-Pak Vac tC18 6 ccm, 500 mg sorbent material).
5. Suitable SPE equipment (Waters Vacuum Extraction Manifold, 20-position + vacuum pump).
6. Lyophilizing equipment.

### 2.5 Phosphopeptide Enrichment with $TiO_2$

1. Wash buffer A. 80% ACN, 0.4% TFA in water.
2. Wash buffer B. 30% lactic acid in Wash buffer A.
3. Elution buffer 1. 1.5% ammonium hydroxide in water.
4. Elution buffer 2. 5% pyrrolidine in water.
5. Acidifying reagent. 2.5% TFA in water.
6. Spin tips with prefilled $TiO_2$ (e.g., from GL Sciences), suitable adaptors, and collection tubes.
7. Benchtop centrifuge.

### 2.6 Desalting II

1. Graphite Resin Spin Columns (e.g., from Thermo Scientific).
2. ACN.
3. 1% TFA in water.
4. 0.1% TFA in water.
5. 1.5% ammonium hydroxide in water.
6. 0.1% formic acid (FA), 50% ACN in water.

### 2.7 MS Analysis

1. A nanoUPLC system coupled to a high-resolution mass spectrometer and their operating and processing software (e.g., nanoAcquity UPLC, Synapt G2-S, and MassLynx/ProteinLynx Global Server (PLGS) from Waters, IsoQuant (available

at www.isoquant.net) and PEAKS Studio from Bioinformatics Solutions Inc.).

2. Appropriate vials for sample injection.
3. Mobile phases A and B, depending on the chromatographic system used.
4. 0.1% FA for sample resuspension and injection.
5. Statistical analysis tools (e.g., Perseus, R, or Python).
6. Protein database in FASTA format for *Magnaporthe oryzae* (e.g., downloaded from www.uniprot.org).

## 3 Methods

A global overview of the proteomics and phosphoproteomics workflow is presented in Fig. 1.

All procedures can be carried out at room temperature, unless otherwise specified.

### 3.1 Lysis of Mycelium

1. Lyophilize filtered cell suspension.
2. Precool mortar and pestle with liquid nitrogen and grind the lyophilized cell material.
3. Prewarm the lysis buffer up to 95 °C in the water bath.
4. Weigh (approx. 200 mg) grinded fungal material into a 5 mL tube.
5. Transfer a nominal volume of 2.5 mL of lysis buffer into the tubes and immediately resuspend (e.g., by re-aspiration) (*see* **Note 1**).
6. Heat the sample for 5 min in the water bath at 95 °C and let it cool 5 min on room temperature. Repeat this step twice.
7. After cooling down to room temperature, add 160 μL alkylation reagent and incubate for 30 min in the dark.
8. Centrifuge the sample at room temperature with 4600 × *g*.

### 3.2 Protein Precipitation, Resuspension, and Quantification

1. Precool centrifuge to 4 °C.
2. Transfer supernatant to a 50 mL Protein LoBind conical tube (*see* **Note 2**).
3. Add 10 mL of methanol and 2.5 mL of chloroform and mix thoroughly (*see* **Note 3**).
4. Add 7.5 mL of water and mix thoroughly.
5. Centrifuge at 4 °C with 4600 × *g* for 1 min.
6. Gently remove the supernatant, leaving the protein interphase intact.
7. Add 10 mL of methanol and mix thoroughly.

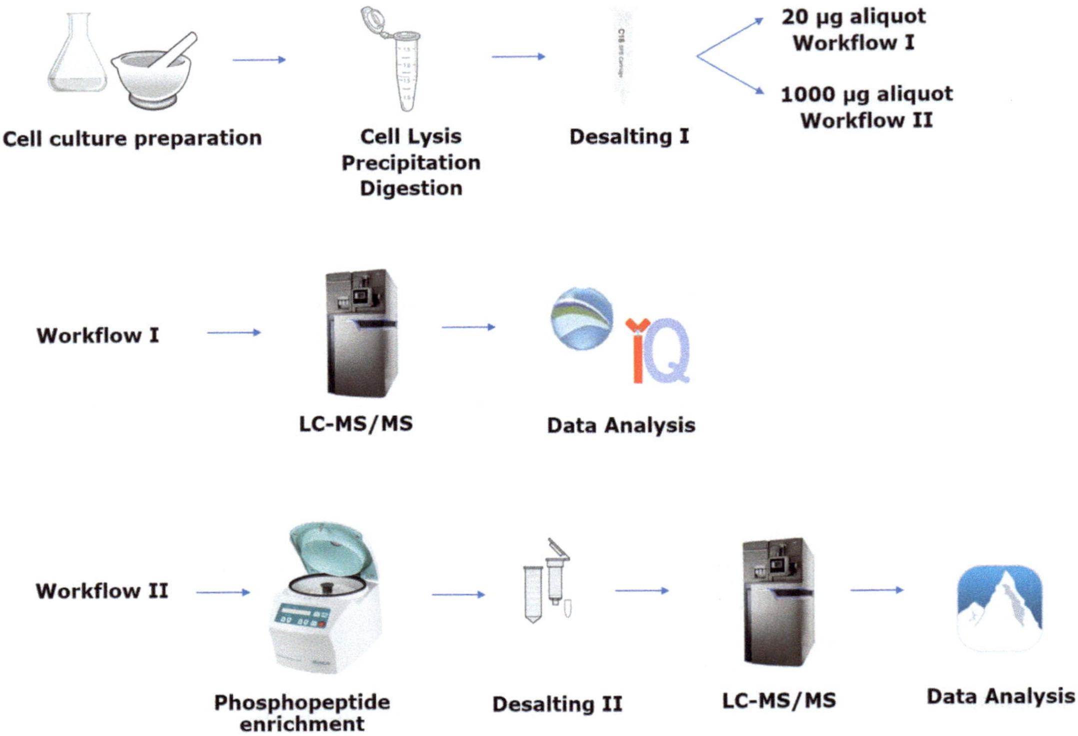


**Fig. 1** Global proteomics and phosphoproteomics workflow overview. First, proteins are proteolytically digested with trypsin. Resulting peptides are aliquoted into 20 μg for global proteome analysis (workflow I) and 1000 μg for phosphoproteome analysis (workflow II). In global proteome analysis, the data is acquired with LC-MS/MS in data-independent-acquisition (DIA) mode. Database search is performed with PLGS (Waters) followed by label-free quantification post-processing with IsoQuant. The phosphoproteome workflow requires an additional enrichment step for phosphopeptides, followed by data acquisition with LC-MS/MS in data-dependent-acquisition (DDA) mode. Raw data processing and label-free quantification are done using PEAKS Studio

8. Centrifuge at 4 °C with 4600 × *g* for 2 min.
9. Gently remove supernatant from the precipitated proteins and let it dry on air for approx. 30 min.
10. Add 2.5 mL of resuspension buffer and mix gently. If necessary, sonicate for 5 min and let proteins dissolve for approx. 30 min (*see* **Note 4**).
11. Sample necessary volume for protein quantification and perform the assay (*see* **Note 5**).

### 3.3 In-Solution Tryptic Digestion

1. Dilute the resuspended proteins 1:4.44 to reduce the urea concentration down to 1.6 M.
2. Add trypsin 1:25 w/w according to the results of the protein quantification.
3. Incubate at 32 °C overnight.

4. After complete digestion, acidify sample by adding 0.5% TFA and mixing. The sample should be foaming due to its AMBIC content.

### 3.4 Desalting I

Always use 2/3 of the cartridge volume available. Here we present the workflow using Waters Sep-Pak tC18 500 mg sorbent/6 $cm^3$ volume. Do not let the sorbent bed dry completely during vacuum application (*see* **Note 6**).

1. Wash the C18 material by flushing with 2/3 cartridge volume methanol (with 500 mg/6 ccm cartridges, i.e., 4 mL).
2. Condition with 50 ACN/0.1% TFA (4 mL).
3. Equilibrate with 0.1% TFA (4 mL).
4. Load samples as slowly as possible (*see* **Note 7**).
5. Wash with 0.1% TFA (4 mL); repeat 5×.
6. Elute samples with 50% ACN/0.1% TFA (4 mL) into appropriate tubes for the following lyophilization. Make sure to mix the complete eluate thoroughly.
7. Aliquot 1: Transfer an appropriate amount for whole proteome analysis to a new tube, typically 20 μg.
8. Aliquot 2: Transfer an amount of 1 mg protein to a new tube.
9. Remaining volume can be stored at −80 °C for later analysis.
10. Lyophilize Aliquot 1 and Aliquot 2 by first freezing and subsequent treatment under high vacuum, ideally in a centrifuge.

### 3.5 Phosphopeptide Enrichment with $TiO_2$

After each step, inspect the tips visually for complete elution or disintegration of the $TiO_2$ material. Refer to Notes for troubleshooting. The following procedure is adopted from the instructions of the Pierce $TiO_2$ Phosphopeptide Enrichment and Clean-up Kit.

1. Resuspend the lyophilized Aliquot 2 in 150 μL of Wash buffer B.
2. Assemble collection tube, adaptor, and $TiO_2$ spin tips in a benchtop centrifuge.
3. Add 20 μL of Wash buffer A and centrifuge at 3000 × *g* for 2 min.
4. Inspect each tip visually if Wash buffer A eluted completely and add 20 μL of Wash buffer B. Centrifuge at 3000 × *g* for 2 min.
5. Place spin tips and adaptor in a new collection tube. Centrifuge resuspended sample and apply to the tip and centrifuge at 1000 × *g* for 10 min (*see* **Note 8**).
6. Reapply the sample and centrifuge again at 1000 × *g* for 10 min.

7. Wash spin tip by adding 20 μL of Wash buffer B and centrifuge at 3000 × *g* for 2 min (*see* **Note 9**).
8. Wash spin tip by adding 20 μL of Wash buffer A and centrifuge at 3000 × *g* for 2 min. Repeat this step twice.
9. Place spin tips and adaptor in a new collection tube. Add 50 μL of Elution buffer 1 and centrifuge at 1000 × *g* for 5 min.
10. Add 50 μL of Elution buffer 2 and centrifuge at 1000 × *g* for 5 min.
11. Remove spin tips and adaptor and add 100 μL of acidifying reagent to the eluate.

### 3.6 Desalting II

If 1.5 mL tubes are used, the collection tubes can hold up to 400 μL, before the flow-throughs from washing steps must be discarded.

1. Remove lid and bottom cap from the graphite spin columns and place in a suitable collection tube.
2. Centrifuge at 2000 × *g* for 1 min. Discard the flow-through.
3. Add 100 μL of 1.5% ammonium hydroxide to the graphite resin, shake gently to suspend the graphite, and centrifuge at 2000 × *g* for 1 min. Repeat once.
4. Add 100 μL of ACN for graphite activation, shake gently, and centrifuge at 2000 × *g* for 1 min.
5. Add 100 μL of 1% TFA for equilibration and centrifuge at 2000 × *g* for 1 min. Repeat once.
6. Transfer the graphite spin columns to a new collection tube. Make sure to homogenize the sample and apply to the graphite bed.
7. Mix gently every 2 min for 10 min to allow binding of the peptides to the material.
8. Centrifuge at 1000 × *g* for 3 min and discard the flow-through.
9. Add 200 μL of 1% TFA and centrifuge at 2000 × *g* for 1 min. Repeat once.
10. Transfer the graphite spin columns to a new collection tube.
11. Add 100 μL of 0.1% FA/50% ACN for peptide elution. Centrifuge at 2000 × *g* for 2 min and repeat this step three times. Collect all four elution fractions in one tube.
12. For lyophilization, freeze the sample at −80 °C and evaporate the solvent with vacuum.

### 3.7 MS Analysis

For MS analysis we would like to refer to the previously published methods, as the in-depth description of instrument settings and procedures would fill another chapter.

1. Dissolve the lyophilized peptides Aliquot 1 and Aliquot 2 in an appropriate amount of 0.1% FA in water, typically 20–40 μL.
2. For nanoUPLC settings, we would like to refer to the previously published methods such as [4]. Both aliquots can be separated with the same chromatographic conditions.
3. We advise to analyze Aliquot 1 (whole proteome) in data-independent-acquisition mode (DIA) as described in [4].
4. For Aliquot 2 (phospho-proteome) we advise to use the data-dependent-acquisition mode (DDA). In short, the suggested settings were as follows: Instrument resolution was operated in "sensitivity mode" at 20.000 (FWHM) resolution. MS1 survey scan time was 0.1 s, and the top 10 highest abundant peptides were selected for fragmentation in the collision cell using *m/z*-dependent collision energies. Scan time for fragment ions was 0.15 s. Dynamic exclusion of fragmented precursor ions was enabled and the exclusion time window was set to 15 s.
5. DIA data are processed in PLGS using the desired *M. oryzae* protein database. Label-free quantification (LFQ) is subsequently performed using IsoQuant software. Suggested settings for protein identification include 15 ppm precursor mass error, 0.03 Da fragment ion error, and 1% false discovery rate (FDR) at peptide level.
6. Phosphopeptide DDA data is processed with PEAKS Studio against the desired *M. oryzae* protein database. LFQ is performed using the PeaksQ algorithm. Suggested settings for phosphopeptide identification include 15 ppm precursor mass error, 0.03 Da fragment ion error, 1% false discovery rate (FDR) at peptide spectrum match (PSM) level, and AScore for identified phosphosites above 20.

## 4 Notes

1. Although the pipetted volume will not be 2.5 mL, due to the almost boiling lysis buffer, we found the optimal weight/volume ratio to be 2.5 mL boiling lysis buffer per 200 mg starting material.
2. If other labware is used, check for chloroform compatibility of the desired labware.
3. For processing large cohorts of samples, it is handy to work with graduated glass pipettes and equipment like manual propipetters (Howorka-Ball, Aspirette). They allow a faster dispensing and aspiration as electronic devices or Multipettes would do, especially when volume accuracy is not necessary.

4. The protein pellet should never dry completely. The presence of large amounts of methanol can be identified via its scent. Proceed with sample preparation, when the methanol scent is unobtrusive.
5. Typically, we sample 10 μL of protein and add 150 μL of quantification reagent in a 96-well ELISA plate. Measurements are done using a TECAN plate reader with filter slides for 660 nm.
6. We identified a 1/100 ratio of peptides/C18 sorbent as optimal. Although there is also a 3 $cm^3$ variant available for the tC18 material offered by Waters, we tend to use the 6 ccm cartridges. These do not clog and are easier to process, if sample centrifugation was incomplete (e.g., lower rpm in the centrifuge).
7. When applying larger volumes, we typically try to adjust the flow to 1–2 drops per second. If less volume is used for lysis and digestion, application speed can be higher and sample can be reapplied to the column.
8. If the sample is difficult to centrifuge, we usually gradually increase the centrifugation speed: first, 10 min at 3000 × *g*, then 10 min at 6000 × *g*, and finally 10 min at 9000 × *g*. We found that increasing the centrifugation speed above 9000 × *g* will cause the disintegration of the $TiO_2$ tip.
9. If the nature of the recovered non-phosphorylated peptides is highly acidic, we recommend to increase the amount TFA in Wash buffer B, instead of increasing the number of washes. In our experiments, the use of three washing steps with Wash buffer B decreased the overall number of recovered peptides dramatically. Furthermore, we found that the presence of DNA eventually reduces the number of identified phosphopeptides. In this case, the use of nucleases before tryptic digest is advised.

## References

1. Altelaar AFM, Munoz J, Heck AJR (2013) Next-generation proteomics: towards an integrative view of proteome dynamics. Nat Rev Genet 14:35–48
2. Garrels JI (1983) Quantitative two-dimensional gel electrophoresis of proteins. Methods Enzymol 100:411–423
3. Abdallah C, Dumas-Gaudot E, Renaut J, Sergeant K (2012) Gel-based and gel-free quantitative proteomics approaches at a glance. Int J Plant Genomics 2012:494572
4. Distler U, Kuharev J, Navarro P, Tenzer S (2016) Label-free quantification in ion mobility–enhanced data-independent acquisition proteomics. Nat Protoc 11:795–812
5. Arike L, Peil L (2014) Spectral counting label-free proteomics. Humana Press, New York, NY, pp 213–222
6. Mann M, Ong S-E, Grønborg M et al (2002) Analysis of protein phosphorylation using mass spectrometry: deciphering the phosphoproteome. Trends Biotechnol 20:261–268
7. Solari FA, Dell'Aica M, Sickmann A, Zahedi RP (2015) Why phosphoproteomics is still a challenge. Mol BioSyst 11:1487–1493
8. Zarei M, Sprenger A, Metzger F et al (2011) Comparison of ERLIC-$TiO_2$, HILIC-$TiO_2$,

and SCX-$TiO_2$ for global phosphoproteomics approaches. J Proteome Res 10:3474–3483

9. Ritorto MS, Cook K, Tyagi K et al (2013) Hydrophilic strong anion exchange (hSAX) chromatography for highly orthogonal peptide separation of complex proteomes. J Proteome Res 12:2449–2457

10. Larsen MR, Thingholm TE, Jensen ON et al (2005) Highly selective enrichment of phosphorylated peptides from peptide mixtures using titanium dioxide microcolumns. Mol Cell Proteomics 4:873–886

11. Wang Y, Wu J, Park ZY et al (2011) Comparative secretome investigation of *Magnaporthe oryzae* proteins responsive to nitrogen starvation. J Proteome Res 10:3136–3148

12. Li Y, Ye Z, Nie Y et al (2015) Comparative phosphoproteome analysis of *Magnaporthe oryzae*-responsive proteins in susceptible and resistant rice cultivars. J Proteome 115:66–80

13. Franck WL, Gokce E, Oh Y et al (2013) Temporal analysis of the *Magnaporthe oryzae* proteome during conidial germination and cyclic AMP (cAMP)-mediated appressorium formation. Mol Cell Proteomics 12:2249–2265

14. Oh Y, Franck WL, Dean RA (2018) Sequential phosphopeptide enrichment for phosphoproteome analysis of filamentous fungi: a test case using *Magnaporthe oryzae*. Methods Mol Biol 1848:81–91

# Chapter 10

# A Target-Based In Vivo Test System to Identify Novel Fungicides with Mode of Action in the HOG Pathway

**Stefan Bohnert, Hendrik Neumann, and Stefan Jacob**


## Abstract

Resistance management plays a key role in modern plant protection. There is a growing need to identify new fungicide targets and new modes of action. In this context, it is also mandatory to find new compounds acting on successful target locations. For the latter, so-called target-site-specific test systems emerged to search for inhibitors. Most of them are based on in vitro assays, in which interaction between a compound and a purified target protein is demonstrated. Consequently, getting essential information about potentially toxic effects in the living cell or in the whole organism is not possible. Thus, we present a fluorescent-labelled mutant strain of the rice blast fungus *Magnaporthe oryzae* as a rapid tool for fluorescence-based identification and visualization of fungicides in vivo with the mode of action in the high osmolarity glycerol (HOG)-signaling pathway. The HOG pathway represents an excellent target for antifungal agents such as the phenylpyrrole fungicides, since almost no relevant resistances have occurred to date, despite 30 years of extensive usage of this fungicide class.


**Key words** *Magnaporthe oryzae*, Fungicide target, Screening system, Target based, In vivo test system, Fludioxonil, High-osmolarity glycerol (HOG) signaling pathway

## 1 Introduction

Fungicides are indispensable to global food security and their use is forecasted to be intensified [1]. Since there is a growing need to produce more calories for the rising global population, the protection of our crops from fungal disease has become of central importance [2]. A major problem is the increasing threat of resistance development, leading to a pressing need to identify novel fungicides or even novel fungicide targets.

One of the most successful drug targets within fungal pathogens is the high osmolarity glycerol (HOG) pathway, which is known to be responsible for osmoregulation in fungi and accepted as the target of phenylpyrrole fungicides, such as fludioxonil [3]. Treatment of the fungus with fludioxonil causes an uncontrolled hyperactivation of this pathway and cell death [4–6]. The

Stefan Jacob (ed.), *Magnaporthe oryzae: Methods and Protocols*, Methods in Molecular Biology, vol. 2356, https://doi.org/10.1007/978-1-0716-1613-0_10, 

osmoregulation system represents an exceptional and flexible fungicide target, since at least two different sensor histidine kinases of the HOG pathway have already been described as fungicide targets in phytopathogenic fungi [7, 8]. In yeast, the translocation of Hog1-type mitogen-activated protein kinases (MAPKs) from the cytosol into the cell nucleus is an immediate response to the activation of the HOG signaling pathway, for example, due to osmotic stress [9–11]. In the pathogenic fungi *Candida albicans*, *Botrytis cinerea*, and *Magnaporthe oryzae*, it was found that the Hog1-type MAPK also migrates into the cell nucleus shortly after osmotic stress or even upon fungicide treatment [2, 3, 12, 13]. Consequently, we decided to use this rapid change of the subcellular localization of the MAPK MoHog1p in *Magnaporthe oryzae* as an indication for pathway activation in a target-based in vivo test system for antifungal compounds. In this protocol, genetic fusion of MoHog1p to the green fluorescent protein (GFP) in the *M. oryzae* wildtype strain represents a powerful tool to visualize the dynamics of osmoregulation and fungicide action. We can refer the mode of action of antifungal compounds directly to the HOG pathway within a few seconds utilizing this method. However, that is only possible because mutants with inactivated HOG signaling are resistant to fungicides having the target located in the HOG pathway [6, 14]. Based on this knowledge, validation of pathway specificity can be assumed by using the "loss of function" (lof) mutant strains *ΔMosln1*, *ΔMohik1*, *ΔMossk1*, *ΔMossk2*, and *ΔMopbs2* producing fluorescence-labeled MoHog1p. The GFP-tagged MoHog1p rapidly translocates into the nucleus upon fungicide treatment when expressed in the wildtype strain, but does not do so when it is expressed in the lof mutants. We also set up the assay with the fungicidal compound geneticin as a further control and to exclude unspecific nuclear translocation of MoHog1p as a result of cell stress. This antibiotic has a completely different mode of action [15]. Treatment with geneticin does not result in an accumulation of GFP fluorescence signals within the nucleus (fluorescence can be observed throughout the cell).

## 2 Materials

1. *Magnaporthe oryzae* cultures, obtained from filter paper disks containing fungal mycelium. Stocks must be dehydrated in desiccator and stored in dry sterile tubes at −20 °C. The *MoHog1p::GFP*-expressing *Magnaporthe* strain and the *MoHog1p::GFP*-expressing lof mutant strains *ΔMosln1*, *ΔMohik1*, *ΔMossk1*, *ΔMossk2*, and *ΔMopbs2* can be used in the assays. Examples of the strategies used to generate the mutants are given for *ΔMohik1/MoHog1::GFP* in [2] (*see* **Note 1**). The lof mutant strain *ΔMohog1* has to be used in the prescreening [16].

2. Double distilled water ($H_2O_{UF}$, 18.2 MΩ).
3. Spatula.
4. Reaction tubes (e.g., 1.5 and 50 mL).
5. Funnel.
6. Filter tissue (e.g., Miracloth).
7. Microtiter plate (96-well).
8. Glass slides with coverslips or chambered coverslips (e.g., 8 wells; 300μL).
9. Fludioxonil stock solution in EtOH; geneticin stock solution in $H_2O_{UF}$.
10. NaCl and/or KCl solution in $H_2O_{UF}$.
11. Vortex mixer.
12. Thermomixer.
13. Inverted microscope for microtiter plate examination.
14. Inverted (for time-course measurement) or upright fluorescence microscope (for counting) (e.g., Axio Imager M2 from Zeiss; with corresponding filter sets for GFP or YFP or red fluorescent protein (RFP) microscopy, e.g., filter set 38 HE ("GFP") EX BP 470/40, FT 495, EM BP 525/50, from Zeiss).
15. Low melt agarose (1.8% w/v) (e.g., from Biozym), only necessary if time-course analysis is intended.
16. Complete medium (CM), contains per liter: 10 g glucose, 1 g yeast extract, 2 g peptone, 1 g casamino acid, 50 mL nitrate salt solution (containing per liter $H_2O$: 120 g $NaNO_3$, 10.4 g KCl, 30.4 g $KH_2PO_4$, 10.4 g $MgSO_4 \times 7H_2O$), and 1 mL of a trace element solution (containing per liter $H_2O$: 22 g $ZnSO_4 \times 7H_2O$, 11 g $H_3BO_3$, 5 g $MnCl_2 \times 4H_2O$, 5 g $FeSO_4 \times 7H_2O$, 1.7 g $CoCl_2 \times 6H_2O$, 1.6 g $CuSO_4 \times 5H_2O$, 1.5 g $Na_2MoO_4 \times 2H_2O$, 50 g $Na_2EDTA$, pH 6.5 adjusted by 1 mol/L KOH).

## 3 Methods

### 3.1 Prescreening

Antifungal compounds are prescreened in 96-well plates by comparing minimal inhibitory concentrations of a test compound on germination of *MoWT* and *ΔMohog1*. Fludioxonil can be used as a positive control, as it inhibits germination in the wildtype strain, whereas it is not active against the mutant *ΔMohog1*.

1. Harvest conidia from 6- to 11-day-old *MoWT* and *ΔMohog1* strains (grown at 26 °C on CM, light/dark 16/8 h) by using $H_2O_{UF}$ and a spatula.

2. Filter conidia through two layers of Miracloth tissue (eventually with a funnel) into a 50 mL reaction tube.
3. Adjust the concentration of conidial suspension to $5 \times 10^4$ conidia/mL in $H_2O_{UF}$.
4. Transfer 200μL of conidial suspension into each well of a 96-well microtiter plate.
5. Add 1μL of fludioxonil stock solution (1 mg/mL in ethanol) to *MoWT* conidial suspension and to *ΔMohog1* conidial suspension as a positive control. Add 1μL of ethanol as a negative control (*see* **Notes 2** and **3**).
6. Add extracts or compounds to be tested to conidial suspensions of *MoWT* and *ΔMohog1* in the wells of the microtiter plate (*see* **Note 3**). Test extracts or compounds in a concentration range usually from 0.1 up to 100μg/mL (e.g., 0.1, 0.3, 1, 3, 10, 30, and 100μg/mL).
7. Mix gently by pipetting and incubate microtiter plate assay without shaking at 22–28 °C overnight (>12 h).
8. Examine conidial germination (and the subsequent initial vegetative growth) in the microtiter plate by using an inverted microscope. Define "non-germinated" as conidia with germ tube length < 10μm. The concentration of extracts or compounds resulting in 50% inhibition of conidial germination is defined as $IC_{50}$ (inhibition concentration). Calculate IC values by counting the number of germinated conidia out of 100 for each sample in triplicate.
9. Retest positively screened extracts or compounds and find the exact $IC_{50}$ values for *MoWT* and *ΔMohog1* by refining the concentration range (*see* **Note 4**).
10. If available, an automated readout to increase throughput in the screening system is possible (*see* **Note 5**).

### 3.2 *Target-Based* In Vivo *Test System*

1. Harvest conidia from 6- to 11-day-old *MoWT*, *ΔMohog1*, and the *MoHog1p::GFP*-expressing lof mutant strains *ΔMosln1*, *ΔMohik1*, *ΔMossk1*, *ΔMossk2*, and *ΔMopbs2* (grown at 26 °C on CM, light/dark 16/8 h) by using $H_2O_{UF}$ and a spatula.
2. Filter conidia through two layers of Miracloth tissue into a reaction tube.
3. Use a conidial suspension of $>1 \times 10^5$ conidia/mL in $H_2O_{UF}$ for manual microscopy.
4. Set up 200μL of conidial suspension, 200 μL of conidial suspension containing 0.25 M KCl or NaCl in $H_2O_{UF}$, 200μL of conidial suspension containing 5μg/mL fludioxonil, and 200μL of conidial suspension containing 250μg/mL geneticin as controls. Vortex 200μL of the conidial suspension with compounds or extracts to be tested (*see* **Note 6**).

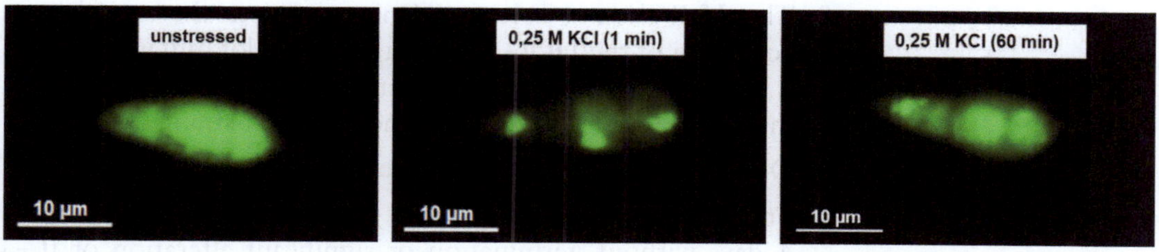


**Fig. 1** Localization of the MoHog1-GFP fusion protein in the mutant strain *MoHOG1::GFP*. Conidia of the mutant strain *MoHOG1::GFP* are shown. Untreated cells and cells after treatment with 0.25 M KCl have been compared. A distinct nuclear accumulation of the GFP signal could be observed 1 min after the KCl stimulus, whereas in unstressed cells and 60 min after the application of KCl, the signal was found throughout the cytosol. (The pictures were taken with the Axio Imager M2 (Zeiss))

5. Transfer in each case 50μL of the conidial suspension (+/− control substances or test substances) on a glass slide and cover carefully with a cover slip.
6. Examine MoHog1p fluorescent protein localization in each mutant manually by using a fluorescence microscope and the corresponding filter set. Identify the nuclear accumulation of fluorescence signal (*see* **Note 7** and also **Note 1**).
7. If time-course analysis is intended, cool the pre-heated low melt agarose to 35 °C by using a thermomixer (with shaking) and mix 1:1 with conidial suspension. Pipette 100μL of that suspension immediately into chambered coverslips. After a resting time of 5 min, record images in the untreated state using a fluorescence microscope (*see* **Note 8**). Subsequently, 100–200μL of the desired compound solution should be added to the suspensions in the wells and time-course measurements or videos can be recorded in order to observe potential compound-induced nuclear accumulation of the fluorescence signal (e.g., *see* Fig. 1).

## 4 Notes

1. An accumulation of the GFP fluorescence signal in the cell nucleus can be observed after a long irradiation of conidia of the *MoHOG1::GFP* mutants with a GFP excitation wavelength, even without fludioxonil stress (unpublished data). Since such a radiation-induced accumulation of the GFP fluorescence signal in the cell nucleus cannot be observed using lower-energy light of the excitation wavelength, i.e., for RFP-mediated radiation, the expression of a *MoHOG1::RFP* construct in corresponding strains would probably be a more robust approach to establish a "time-independent" screening system for HOG pathway-dependent bioactive compounds.

2. The *MoWT* is not able to germinate due to the toxic hyperactivation of the HOG signaling pathway; lof mutants, such as *ΔMohog1*, are vital, because hyperactivation of the HOG signaling pathway is not possible in these mutant strains.
3. Do not exceed 1% (v/v) of solvent in the germination assay.
4. Fresh conidial suspensions can be stored at 4 °C for up to 4 days without germination or significant alteration of $IC_{50}$ values under fludioxonil stress after setting up a new assay. However, this cultivation period could alter $IC_{50}$ values for other test substances.
5. The screening system can be scaled up by expanding the cultivation time of the assay mixtures and using a microtiter plate reader and measuring the optical density (i.e., conidial germination and mycelial growth). However, the assay should be set up by using a complete medium instead of $H_2O$ to ensure a strong vegetative growth of the strains.
6. Do not exceed 0.5% (v/v) of the solvent (ethanol, dimethyl sulfoxide) or 0.1% (v/v) (acetonitrile) as it will lead to false-positive nuclear localization of the MoHog1 fluorescent protein.
7. By using a strain expressing a MoHog1-GFP fusion protein, do not exceed an observation time period (with permanent UV radiation of conidia) of 5 min (depending on the radiation strength of the UV lamp) to avoid false-positive nuclear accumulation; pre-test a critical time period in your setup by using conidial suspension in $H_2O_{UF}$ without induction by extracts or compounds (*see* **Note 1**).
8. Using standard agarose results in a false-positive nuclear localization of MoHog1 fluorescent protein because of heat.

## References


1. Zubrod JP, Bundschuh M, Arts G et al (2019) Fungicides: an overlooked pesticide class? Environ Sci Technol 53:3347–3365
2. Bohnert S, Neumann H, Thines E, Jacob S (2019) Visualizing fungicide action: an *in vivo* tool for rapid validation of fungicides with target location HOG pathway. Pest Manag Sci 75:772–778
3. Kojima K, Takano Y, Yoshimi A et al (2004) Fungicide activity through activation of a fungal signalling pathway. Mol Microbiol 53:1785–1796
4. Bohnert S, Heck L, Gruber C et al (2018) Fungicide resistance towards fludioxonil conferred by overexpression of the phosphatase gene *MoPTP2* in *Magnaporthe oryzae*. Mol Microbiol 111:mmi.14179
5. Motoyama T, Ohira T, Kadokura K et al (2005) An Os-1 family histidine kinase from a filamentous fungus confers fungicide-sensitivity to yeast. Curr Genet 47:298–306
6. Jacob S, Foster AJ, Yemelin A, Thines E (2015) High osmolarity glycerol (HOG) signalling in *Magnaporthe oryzae*: identification of *MoYPD1* and its role in osmoregulation, fungicide action, and pathogenicity. Fungal Biol 119:580–594
7. Yoshimi A, Kojima K, Takano Y, Tanaka C (2005) Group III histidine kinase is a positive regulator of Hog1-type mitogen-activated protein kinase in filamentous fungi. Eukaryot Cell 4:1820–1828
8. Jacob S, Schüffler A, Thines E (2016) Hog1p activation by marasmic acid through inhibition

of the histidine kinase Sln1p. Pest Manag Sci 72:1268–1274

9. Ferrigno P, Posas F, Koepp D et al (1998) Regulated nucleo/cytoplasmic exchange of HOG1 MAPK requires the importin beta homologs NMD5 and XPO1. EMBO J 17:5606–5614
10. Van Wuytswinkel O, Reiser V, Siderius M et al (2000) Response of *Saccharomyces cerevisiae* to severe osmotic stress: evidence for a novel activation mechanism of the HOG MAP kinase pathway. Mol Microbiol 37:382–397
11. O'Rourke SM, Herskowitz I, O'Shea EK (2002) Yeast go the whole HOG for the hyperosmotic response. Trends Genet 18:405–412
12. Cheetham J, Smith DA, da Silva DA et al (2007) A single MAPKKK regulates the Hog1 MAPK pathway in the pathogenic fungus *Candida albicans*. Mol Biol Cell 18:4603–4614
13. Heller J, Ruhnke N, Espino JJ et al (2012) The mitogen-activated protein kinase BcSak1 of *Botrytis cinerea* is required for pathogenic development and has broad regulatory functions beyond stress response. Mol Plant-Microbe Interact 25:802–816
14. Ochiai N, Fujimura M, Motoyama T et al (2001) Characterization of mutations in the two-component histidine kinase gene that confer fludioxonil resistance and osmotic sensitivity in theos-1 mutants of *Neurospora crassa*. Pest Manag Sci 57:437–442
15. Bar-Nun S, Shneyour Y, Beckmann JS (1983) G-418, an elongation inhibitor of 80 S ribosomes. Biochim Biophys Acta 741:123–127
16. Jacob S, Foster AJ, Yemelin A, Thines E (2014) Histidine kinases mediate differentiation, stress response, and pathogenicity in *Magnaporthe oryzae*. Microbiology 3:668–687

# Chapter 11

# Making Visualization Work for You: Deriving Valuable Insights from Omics Data

Alexander Yemelin

## Abstract

"Omics" technologies (genomics, transcriptomics, proteomics, metabolomics, etc.) have significantly improved our understanding of biological systems. They have become standard tools in biological research, for example, identifying and unraveling transcriptional networks, building predictive models, and discovering candidate biomarkers. The rapid increase of omics data presents both a challenge and great potential when it comes to providing valuable insights into the underlying patterns of the investigated biological processes. The challenge is extracting, processing, integrating, and interpreting the corresponding datasets. The potential, on the other hand, arises from generation of verifiable hypotheses to understand molecular mechanisms behind biological processes, for example, gene expression patterns. Exploratory data analysis techniques are used to get a first impression of the important characteristics of a dataset and to reveal its underlying structure. However, investigators are often faced with the difficulties of managing the high-dimensional nature of the data. In order to efficiently analyze biological data and to gain a deeper understanding of underlying biological mechanisms, it is essential to have robust and interactive data visualization tools.

**Key words** Python, *Bokeh*, Panel framework, *Magnaporthe oryzae*, Interactive visualization, Data exploration, Transcriptomics, RNA-seq analysis, Jupyter notebook, PaperBLAST

## 1 Introduction

The aim of omics data analysis is to translate large amounts of data that cover numerous samples, conditions, and time points into structured, domain-specific knowledge that can guide research decisions. There is a growing need for interactive, graphical representations of omics data. While there are a number of different visualization tools available, their functionality is usually limited to predefined data analysis features not designed with the end user researcher in mind. Additionally, these existing tools offer limited interactivity, thus making it very difficult to perform task-

Stefan Jacob (ed.), *Magnaporthe oryzae: Methods and Protocols*, Methods in Molecular Biology, vol. 2356,
https://doi.org/10.1007/978-1-0716-1613-0_11, 

specific data exploration and knowledge extraction [1–3]. Unfortunately, researchers are faced with either doing all or parts of their data analysis manually or are in some cases forced to limit their data exploration activities. This chapter was motivated by the belief that in order to be effective, researchers need flexible, customizable, and interactive tools designed to address their specific data analysis tasks, instead of trying to adapt their work to fit within the limits of existing tool functionality.

Frameworks for building tailored applications that generate visual representations play a key role in modern data analysis [4]. Python continues to be the language of choice for solving problems in computational biology. With Python, initial data exploration is typically performed in a *Jupyter Notebook*, using tools like *Matplotlib* and *Bokeh* to develop static or interactive plots. These tools support a simple syntax for making certain types of plots, but showing more complex data relationships often requires a more effective data exploration environment, for example, a standalone server environment, running a web app [5]. *Dash* and *Bokeh Server* address these requirements and represent two popular frameworks for developing web-based data dashboards in Python. Within this protocol, we focus on implementing and leveraging of *Bokeh* Server-based visualizations maximizing knowledge acquisition from the biological data.

### 1.1 Why Use Bokeh?

*Bokeh* is a Python visualization library that provides a powerful and flexible declarative framework for creating web-based plots [6]. *Bokeh* renders plots using HTML canvas and includes many compelling interactivity features. For instance, it includes widgets that can work in both Jupyter Notebook and server environments (https://docs.Bokeh.org/en/latest/). *Bokeh* is fundamentally different from other Python libraries like *Matplotlib* and *Seaborn* which render plots as static. It leverages the full power of JavaScript to provide interactive controls like zooming in and dragging and panning out of the visual panel. While this sort of simple functionality can be written using pure Python, *Bokeh* allows additional flexibility to develop custom JavaScript callbacks defining specific interactions that lie outside the scope of the core functionality of the library. It is therefore possible to create powerful dashboards and responsive visualizations that could convey much more information than a static picture ever could. In general, *Bokeh* is well documented and has a very active community that helps users find quick solutions to various technical obstacles. The gallery on the developer's website provides a rich set of examples presenting *Bokeh's* capabilities (*see* **Note 1**). This chapter will get you excited about the possibilities of using *Bokeh* to build powerful and interactive visualization tools for exploration of complex omics data.

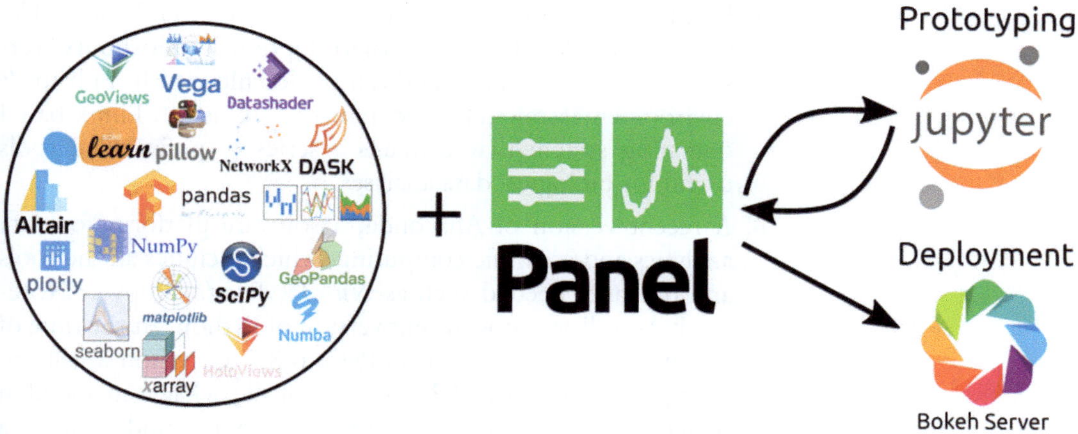


**Fig. 1** Visualization of Panel integration with Python ecosystem (source: https:/medium.com/@philipp.jfr/panel-announcement-2107c2b15f52)

### *1.2 Extend* Bokeh*'s Functionality by Using* Bokeh*/Panel Framework*

*Panel* is a new high-level open-source library that was developed to integrate the outputs of existing tools in the *PyData* ecosystem (Fig. 1) as a dashboard or app.

*Panel* is built on top of *Bokeh* and supports other popular Python plotting libraries for data visualization such as *Matplotlib*, *Plotly*, and *Datashader*. It allows composition of reactive and viewable objects into versatile plots, tables, and other visualizations, linking them all at once into simple, one-off apps to do a specific exploratory task (https://panel.holoviz.org/). Similar to *Bokeh*, *Panel* offers support for Jupyter Notebooks, allowing but not requiring their deployment. Hence, it ensures that users can seamlessly work with already familiar analysis and visualization tools.

In this chapter we provide a step-by-step workflow for generation of an interactive scatterplot for comprehensive data analysis using published RNA-seq experimental data obtained from our previous work with *Magnaporthe oryzae*. Furthermore, we provide guidelines for easy integration of other components into the *Panel* framework, allowing for an integrative approach of data processing. Once developed, the plot displayed on the web page is aesthetically pleasing and can be easily manipulated within the browser. Its functionality supports key research tasks, having been designed from the start with the end user in mind.

## 2 Materials and Resources

1. Personal computer with Internet access and minimal hardware requirements. For the work described in this protocol, we used 8 GB RAM, 4 processor cores, and 1 TB of free disk space. Computer requirements are dependent on the amount of data to be analyzed.

2. Windows, Mac OS, and Linux with installed web browser like Chrome, Firefox, Opera, or Safari. Optionally Bio-Linux (version 8) is recommended and can be downloaded from http://environmentalomics.org/bio-Linux/. It is a Linux-based operating system that contains a series of preinstalled tools useful for biological data analyses.
3. A recent version of Anaconda, a software platform for data analytics and scientific computing, which includes all the tools and packages needed such as *NumPy*, *Pandas*, Jupyter Notebook, as well as a most recent version of Python. At the time of writing, users should start on Python 3.7 as Python 2 will not be supported beyond 2020. The list of all packages included in Anaconda is available at https://docs.anaconda.com/anaconda/packages/pkg-docs.
4. Some basic knowledge of programming using Python and some experience with Git would be beneficial. Additionally, the reader is advised to become familiar with the basics of Python and Git before reading this chapter. Pittard and Li provide a very good written guide on using Python, R, and Git for biologically oriented researchers [7].
5. Gene expression information obtained from RNA-seq data. This data may be found in public databases or obtained in-house. Here we use RNA-seq information from previous study with *Magnaporthe oryzae* [8].

## 3 Methods

Scatterplots represent a commonly used way to display results of differential expression analyses or other omics experiments. In particular, they are used to evaluate the relationship between two different continuous variables at once. To generate the most basic scatterplot of RNA-seq results, we need a file of differentially expressed results. The file used here was generated using DESeq2 from a previous study [8] and can be downloaded from the GitHub link provided. Any other output file from any RNA-seq tool for differential expression analysis, such as edgeR, can be used as long as the required normalized expression values (e.g., FPKM, **f**ragments **p**er **k**ilobase of exon model per **m**illion reads mapped) are calculated (Fig. 2). However, the user should note that methods, tools, and parameters used to analyze a given experiment should be determined on a case-by-case basis (*see* **Notes 2** and **3**).

### 3.1 Data Files Used

In the following analysis, we used a subset of the data measured by Bohnert et al. [8]. The authors provide a genome-wide transcriptomic analysis of *M. oryzae* strains before and after 25 min salt stress (0.5 M KCl), thereby presenting major differences at the

| | A | B | C | D | E | F | G | H |
|---|---|---|---|---|---|---|---|---|
| 1 | AlternateID | Annotation | wt_0min | wt_25min | dhog_0min | dhog_25min | dhoga_0min | dhoga_25min |
| 2 | MGG_00019 | galactose-proton symport | 51 | 29.7 | 3.7 | 4.7 | 0 | 0 |
| 3 | MGG_00040 | high-affinity glucose transporter ght2 [Magnaporthe oryzae 70-15] | 28.4 | 711.4 | 12 | 16.1 | 32.5 | 41.9 |
| 4 | MGG_00063 | glycogen debranching enzyme | 62.7 | 170.7 | 17.5 | 21.7 | 24.9 | 37.4 |
| 5 | MGG_00081 | cell wall | 119.6 | 523.4 | 33.1 | 28.4 | 67.1 | 51.9 |
| 6 | MGG_00087 | Ribosome biogenesis SSF1 | 9.8 | 6.2 | 56.1 | 66.2 | 30.3 | 30.1 |
| 7 | MGG_00097 | reductase [Fusarium fujikuroi] | 163.7 | 1456.2 | 13.8 | 21.7 | 11.9 | 48.3 |
| 8 | MGG_00099 | phosphatidyl synthase | 602.8 | 3653.6 | 31.3 | 91.7 | 56.3 | 329.8 |
| 9 | MGG_00119 | hypothetical protein MGG_00119 [Magnaporthe oryzae 70-15] | 61.7 | 129.9 | 25.7 | 24.6 | 8.7 | 4.6 |
| 10 | MGG_00126 | kelch domain-containing | 3.9 | 105.2 | 0 | 0 | 1.1 | 0 |
| 11 | MGG_00127 | saccharopine dehydrogenase | 8.8 | 54.4 | 3.7 | 3.8 | 4.3 | 2.7 |
| 12 | MGG_00128 | sporulation RMD1 | 20.6 | 105.2 | 8.3 | 10.4 | 8.7 | 12.8 |

**Fig. 2** A snippet of RNA-seq input data derived from DESeq2 analysis illustrating the required columns for generation of scatterplot application. The values are expressed as FPKM

transcriptome level between WT (70–15), *Δhog1*, and *Δhog1 (adapted)* strains. The differentially expressed genes (DE genes) were identified by DEseq2. Genes with an adjustment of *p*-values by Benjamini-Hochberg FDR (FDR $< 0.05$) and fold change values $>2.5$ were considered to be differentially expressed. For the sake of simplicity, for our visualization we use already pre-filtered DE genes from the publication, thus considering only the FPKM values of the DE genes. However, using the tools presented in this chapter, the interested user can easily implement additional filtering options like sliders, allowing filtering based on the statistics of choice, e.g., *p*-values (*see* **Note 4**).

The common methodology for building a visualization with *Bokeh* involves the following steps:

- Import the library and functions/methods.
- Prepare the data to analyze.
- Set the output mode (where the visualization will be rendered: rendering inline in a Jupyter Notebook, a static HTML file in a web browser, or deployment of a web app on server).
- Set up the figure with different styling features (if required).
- Organize the layout.
- Visualize the plot.

### 3.2 Building a Scatterplot App

After Anaconda has been installed, install *Bokeh* and Biopython with *conda* command:

```
conda install Bokeh, biopython
```

*Bokeh* has two methods for creating visualizations. The first approach is to generate HTML documents that can be used standalone or embedded in a Jupyter Notebook. The second is to build a *Bokeh*-App that provides more flexibility and customization options but requires a separate application to be run to serve the data. In this guide we will start by following the first approach. At

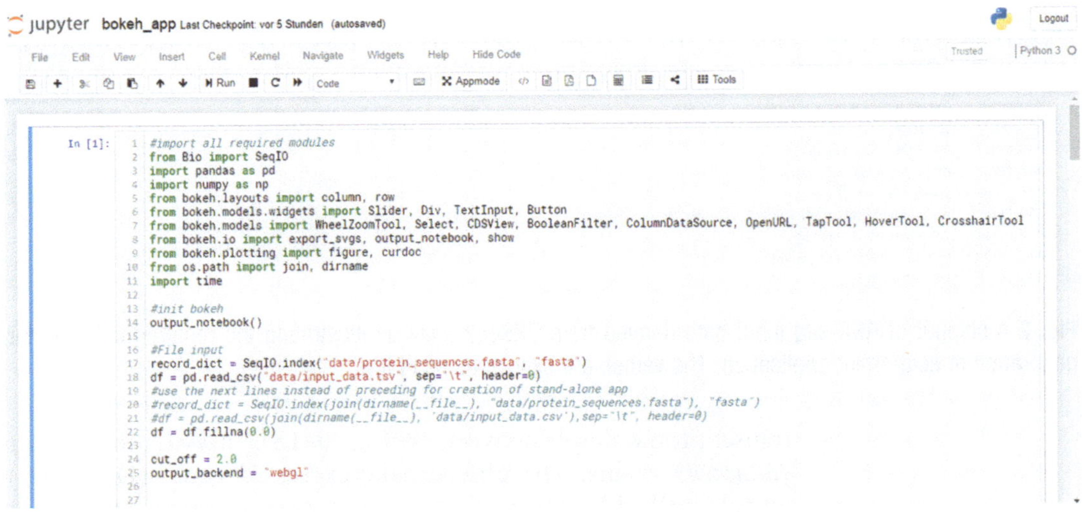


**Fig. 3** An example of a Jupyter Notebook showing a code excerpt of the implemented Bokeh-App

the end of this book chapter, we transform our App to a more advanced version supporting the deployment of different plots in the form of distinct tabs within the single web app. We achieve this transformation by following the second method using the single-folder approach described on the official *Bokeh* website.

The Jupyter Notebook App can be launched by clicking on the Jupyter Notebook icon installed by Anaconda in the Start Menu (Windows) or by typing in a terminal (cmd on Windows):

```
jupyter notebook
```

This will launch a new browser window (or a new tab) showing the Notebook Dashboard, a sort of control panel that allows to create a new notebook or to select an already existing one (Fig. 3).

Create a new folder, e.g., on Desktop, and save the newly created Jupyter Notebook in it. We start writing our code in the first cell of the notebook:

1. The first step is to import several modules needed to build the App:

```
#import all required modules
from Bio import SeqIO
import pandas as pd
import numpy as np
from bokeh.layouts import column, row
from bokeh.models.widgets import Slider, Div, TextInput, Button
from bokeh.models.widgets import WheelZoomTool, Select, CDSView, BooleanFilter, ColumnDataSource, OpenURL, TapTool, HoverTool, CrosshairTool
from bokeh.io import export_svgs, output_notebook, show
from bokeh.plotting import figure, curdoc
```

```
from os.path import join, dirname
import time

#init bokeh
output_notebook()
```

2. The next step is to import the csv file containing the RNA-seq data as a pandas DataFrame allocated to a variable *df*. Variable *url* defines the relative path of the directory where all necessary files will be located later. We also initialize a variable named *record_dict* with the contents of file *protein_sequences.fasta.* This file contains protein sequences for all annotated genes of *M. oryzae* and is required for the implementation of a *tap tool* which will be used for PaperBLAST search algorithm [9].

```
#File input
record_dict = SeqIO.index("data/protein_sequences.fasta", "fasta")
df = pd.read_csv("data/input_data.tsv", sep="\t", header=0)
#replace the above lines with the following to create a standalone app
#record_dict = SeqIO.index(join(dirname(__file__), "data/protein_sequences.fasta"), "fasta")
#df = pd.read_csv(join(dirname(__file__), 'data/input_data.csv'), sep="\t", header=0)
df = df.fillna(0.0)
```

3. Once we initialize the cut-off value for the filtering slider and set *backend_output* to "*webgl*" to improve performance, we can proceed with the declaration of our plot generation function in **step 4**. (The variable *backend_output* will be used later in the context of saving the plot as a high-resolution image; *see* **step 12.**)

```
cut_off = 2.0
output_backend = "webgl"
```

4. Start implementing the function *create_figure()*. It takes two arguments, *cut_off* and *output_backend*, and returns a figure object associated with variable *p*. Within this function we provide all information related to the styling and addition of interactive tools like *tap tool* and *hover tool* used for the creation of the scatterplot.

```
def create_figure(cut_off, df, output_backend):
columns = sorted(df.columns[2:])

#Initialize variables for data input
xs = df[x.value].values
```

```
ys = df[y.value].values
geneid = df['AlternateID'].values
desc = df['Annotation'].values
x_title = x.value.lower()
y_title = y.value.lower()

#create a new column in df named "seq" storing protein sequences
result={}
for i in geneid:
if i in record_dict.keys():
result[i]= str(record_dict[i].seq)
else: result[i] = str('*')
df['seq']=result.values()
seq = df['seq'].values
```

5. *ColumnDataSource* is a fundamental data structure of *Bokeh* that maps string column names to sequences of data. We can create a *ColumnDataSource* object directly from a dictionary by passing the dictionary object to the class constructor. The dictionary object contains all the data needed for our analysis. The required information is extracted as columns from DataFrame *df* and allocated to variables named *xs*, *ys*, *geneid*, *desc*, and *seq*. Variables *x_value* and *y_value* are associated with the sample names for option selection in the drop-down menu widget.

```
data = {'x_values': xs,
'y_values': ys,
 'fold_change': [(-x/y) if (x > y and y!=0) else (y/x) if (y=x and x!=0) else 'NaN' for x,y in zip(xs,ys)],
'GeneID': geneid,
 'Annotation': desc,
'seq': seq
}


mysource = ColumnDataSource(data=data)
```

6. Another important *Bokeh* structure is *CDSView*. It allows data filtering by using a defined filter as a parameter to *CDSView*-object constructor. It can be then passed in as a view argument to the renderer-adding method *p.circle()*.

```
#use CDSView function to determine a subset of significant genes for visualization
significant_xy_view = CDSView(source=mysource,
```

```
filters=[BooleanFilter([True if ((geneX > 0 and geneY > 0
and(abs(geneX/geneY) > cut_off)) or (geneX > 0 and geneY > 0
and abs(geneY/geneX) > cut_off) and (geneX+geneY) > 10))
else False for geneX,geneY in zip(mysource.data['x_values'],
mysource.data['y_values'])])])
```

7. *Bokeh* uses **figure** as the base object for creating a visualization. Once a figure is created, items can be placed on the figure.

```
#create a plot and style its properties
kw = dict()
kw['title'] = "%s vs %s" % (x_title, y_title)

#make a plot
p = figure(plot_height=500, plot_width=700, output_backend=output_
backend, tools='pan,box_zoom,wheel_zoom,reset,save,tap', **kw)

#make sure WheelZoomTool is activated by default
p.toolbar.active_scroll = p.select_one(WheelZoomTool)
p.xaxis.axis_label = x_title
p.yaxis.axis_label = y_title
p.yaxis.axis_label_text_font_size = '10pt'
p.xaxis.axis_label_text_font_size = '10pt'
p.yaxis.major_label_text_font_size = '10pt'
p.xaxis.major_label_text_font_size = '10pt'
p.title.text_font_size = '12pt'

common_circle_kwargs = {
'x': 'x_values',
'y': 'y_values',
'source': mysource,
'hover_color': 'black'
}

p.circle(**common_circle_kwargs,color="skyblue", alpha=0.7, size=9,
line_color="white")

p.circle(**common_circle_kwargs,view= significant_xy_view,
color="red", muted_alpha=0.1,
legend_label='sig. genes', size=9, line_color="black")

#Further filter by string in title if it is provided
if desc_sel.value != "":
selected = df[df.Annotation.str.contains(desc_sel.value,
case=False) == True]
p.circle(x=selected[x.value].values, y=selected[y.value].values, col-
```

```
or="yellow",
alpha=1.0, muted_alpha=0.1, legend_label='sel. genes',
size=8, line_color="black")
```

8. Optionally add some more cosmetic features to the legend and the toolbar:

```
#Set autohide to true to only show the toolbar when mouse is over plot
p.toolbar.autohide = True

p.legend.title = 'Click to hide'
p.legend.location = "bottom_right"
p.legend.border_line_width = 1
p.legend.border_line_color = "black"
p.legend.border_line_alpha = 0.5
```

9. Every *Bokeh* plot supports interactive tools. Here's what the standard tools look like for our App (Fig.4).

   The actual building of the tools is straightforward. You have the option of defining tools by passing a list of strings as a parameter to the *figure* function call, as shown in the code excerpt in step 7:

```
p = figure(... tools='pan,box_zoom,wheel_zoom,reset,save,
tap'...)
```

   When you use this approach, it is not possible to customize the tools. To have more flexibility, define them separately, as shown below. For instance, it is useful to define a *tap tool* to call PaperBLAST and a *hover tool* to display the fold changes and description of selected genes. Use any column of the data that is available to us in our DataFrame and reference it with symbol @.

```
#use the "seq" column of mysource to complete the URL
# e.g. if the glyph at index 1 is selected, then @seq
# will be replaced with mysource.data['seq'][1]
url = "http://papers.genomics.lbl.gov/cgi-bin/litSearch.cgi?query=@
seq&Search=Search"
taptool = p.select(type=TapTool)
taptool.callback = OpenURL(url=url)
```

```
# Format the tooltip
tooltips = [
('GeneID','@GeneID'),
('Annotation','@Annotation{safe}'),
```

**Fig. 4** An example of standard interaction tools in *Bokeh*

```
('fold change','@fold_change'),
('FPKM (%s)' % (x_title),'@x_values'),
('FPKM (%s)' % (y_title),'@y_values')]


#Configure a renderer to be used upon hover
hover_glyph = p.circle(**common_circle_kwargs, size=15, alpha=0, hover_fill_color='black', hover_alpha=0.5)

#Add the custom HoverTool to the figure
p.add_tools(HoverTool(tooltips=tooltips, renderers=[hover_glyph]))

#Add the custom CrosshairTool to the figure
p.add_tools(CrosshairTool(line_width=1))

#Add interactivity to the legend
p.legend.click_policy="hide"

return p
```

10. The basic plot is structured; handle changes to the data and make sure the visualization is updated accordingly. With the addition of *update_data()* function, *Bokeh* keeps the visualization updated. This function will be called each time we select a sample from the drop-down menu or move the slider to define the cut-off value of fold change in order to filter the data as defined in the next step.

```
def update_data(attr, old, new):
cut_off = float(sl.value)
layout.children[1] = create_figure(cut_off, df, output_backend)
```

11. Create a panel of interaction tools such as a drop-down menu, a slider, and a *TextInput* field, used to communicate with the plot interactively. Note that each object calls on change the *update_data()* function declared in the preceding step.

```
#Add control elements
x = Select(title='X-Axis: condition A', value='wt_0min', options=sorted(list(df.iloc[:,2:].columns)))
```

```
x.on_change('value', update_data)

y = Select(title='Y-Axis: condition B', value='dhog_25-min',
 options=sorted(list(df.iloc[:,2:].columns)))
y.on_change('value', update_data)

sl = Slider(start=0.0, end=10.0, value=cut_off, step=0.5,
 title='|fold change| \u003E')
sl.callback_policy='mouseup'
sl.on_change('value_throttled', update_data)

desc_sel = TextInput(title="Gene annotation contains:")
desc_sel.on_change('value', update_data)
div = Div(text="""
 Interact with the widgets below to query a list of significant genes to plot.
Hover over the circles to see more information about each gene.
Click on the circle to get information from PaperBLAST""")
```

12. To activate the SVG backend, set attribute *p.output_backend* to "*svg*". This allows to save the plot as a high-resolution image for further manipulation using image editing programs such as Adobe Illustrator and Photoshop and/or converted to PDFs. Then create a "save as" button interface and link it to the *button_handler()* declared. Note that clicking the button automatically will save the plot picture in the parent directory rather than calling the browser to ask where to save it.

```
#add button for saving plot as .svg
def button_handler():
timestr = time.strftime("%Y%m%d-%H%M%S")
p.output_backend = 'svg'
export_svgs(p, filename="scatterplot{}.svg".format(timestr))
p.output_backend = 'webgl'



button = Button(label="Save as .svg", button_type="success")
button.on_click(button_handler)
```

13. Now the program is almost finished. Create an instance of the plot and store it in a variable named *p*:

```
#create an instance of the plot
p = create_figure(cut_off, df, output_backend)
```

14. Put all elements in one layout and display the resulting plot in the notebook by calling the imported function *show()*. By

clicking on a "play" button in the upper part of the Jupyter Notebook navigation panel, the entire code is executed and the plot is generated within the Jupyter environment. Note that this way of creating interactive plots allows quick implementation within notebook environment supporting a convenient way of debugging. However, the resulting widget hosting plots are not fully interactive, since sliders and drop-down menus are not supported by default and require code changes (*see* **Notes 5–7**). To create a standalone App, continue with **steps 15** and **16** described below.

```
#Put the controls and plot in a layout and display in the next notebook output cell
controls = column( x, y, sl, desc_sel, button, width=200)
layout = row(column(div, controls), p)
curdoc().add_root(layout)
curdoc().title = "DE Gene Analysis - M. oryzae strains"
```

15. For building a standalone App add the created model with its layout to the current document by calling function *curdoc()* and define a title for the browser window:

    In addition make sure to use the commented out version of the code in **step 2**, where we read in the input data by passing the built-in *__file__* attribute as an argument to the *dirname()* function(*see* **Note 8**).

16. Create a folder named *Bokeh_app* and save the script as *main.py* in it. Finally, to execute our *Bokeh* application, navigate to the parent directory of your application folder *Bokeh_app* (e.g., Desktop) and run the following from the command line interface (e.g., Anaconda prompt):

```
Bokeh serve --show Bokeh_app
```

This will set up a *Bokeh* Server and automatically open the application in your browser. We can optionally use --**port 5000** to specify that the *Bokeh* Server should listen on port 5000. By default, port 5006 is used. We can then access the application and explore the data using an interactive scatterplot with implemented features.

The resulting interactive scatterplot is shown below (Fig. 5).

You can try it at the following binder link:

launch binder

```
https://mybinder.org/v2/gh/Alexyem1/bookchapter_example1/master?urlpath=%2Fproxy%2F5006%2FBokeh_app
```

The *Bokeh* Server will start and load the scatterplot at the address ***http://localhost:5006/Bokeh_app***. Since the plot is

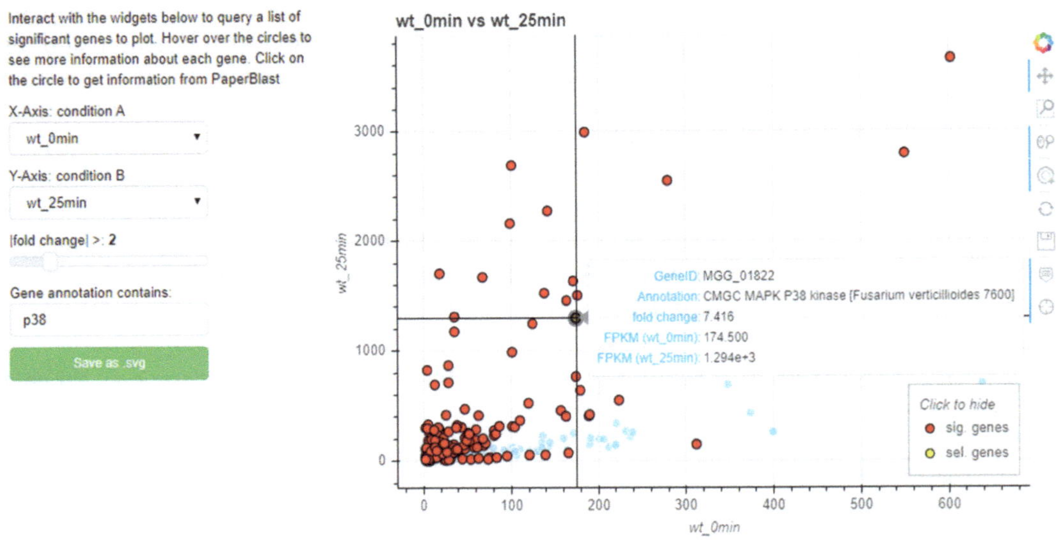


**Fig. 5** *Bokeh* scatterplot visualization of transcript differential expression from the experiment measuring gene expression in different strains of *M. oryzae*. The scatterplot shows the transcripts expressed at levels defined by the control slider implemented. Hovering over a data point of choice shows its label and data; columns for control and treatment FPKM are visualized, and gene name is used for labeling. In the example, HOG1 gene shows highly increased expression in the treatment condition (25 min salt stress) as compared to the control condition (0 min salt stress)

interactive, you can use the *zoom* button on the top right of the plot and select the area you would like to zoom into by drawing a rectangle.

You can hover over points to see the tooltips we added with the hover argument. You can also play around with other interactive components such as panning and zooming.

In the plot above the genes are colored red if they pass the threshold for the fold change values set by the control slider. We can also label one or more genes of interest using the text input widget. This enables us to visualize where these genes are in terms of significance and in comparison to the other genes. Furthermore, using the implemented *TextInput* field, we can search for a specific term (even part of a word will be recognized as a substring) contained in the specified column Annotation for all genes of the input file *input_data.csv*. For instance, looking for "*p38*" results in one hit corresponding to the Hog1p homolog in *F. oxysporum* which is colored yellow in the scatterplot. By clicking on *sel.genes* in the legend, we can hide or activate the highlighting of the retrieved genes. Making use of the *tap tool* we can get additional valuable information for the gene. Clicking on a circle opens a new browser tab which links to PaperBLAST tool [9], showing literature search results related to the corresponding protein sequence (Fig. 6).

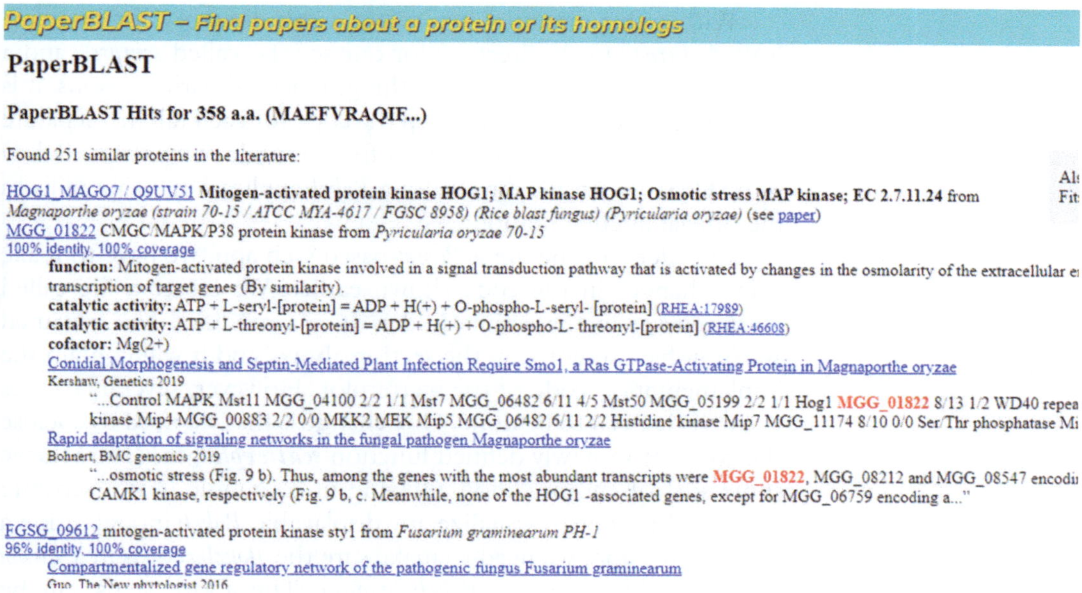


**Fig. 6** Example of PaperBLAST results for *HOG1* protein sequence. PaperBLAST delivers a list of articles and up to two snippets for each protein that is similar to the query protein sequence

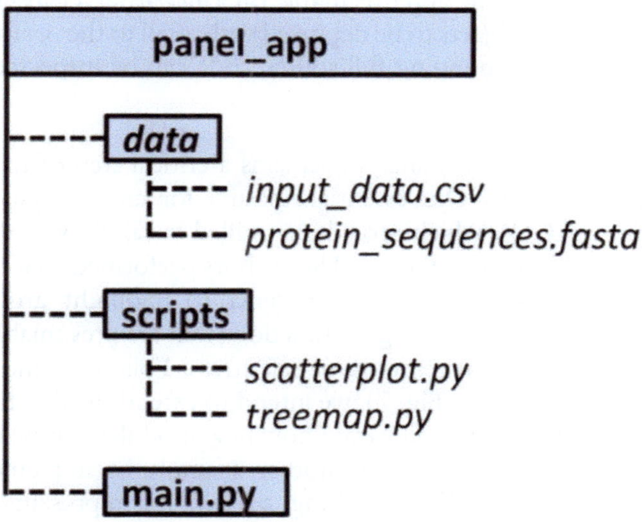


**Fig. 7** Structure of filesystem directory behind the implementation of the *Panel*-based *Bokeh*-App

### 3.3 Building an Advanced Panel-Based Web App

As mentioned earlier, *Bokeh* applications may be created by maintaining and populating a filesystem directory with appropriate files. To start a (*Panel* based) *Bokeh* application, change directory to its parent directory and execute ***Bokeh serve --show*** with the corresponding name of the application directory. Hence, to set up a *Panel*-based *Bokeh* application, we can create one parent directory (e.g., on the Desktop) to hold everything called *panel_app* (Fig. 7).

Within this directory, we create a sub-directory for our data (called *data*), a sub-directory for our scripts (called *scripts*), and a *main.py* script to combine everything. For modularity reasons, it is generally a good practice to keep the code for each tab in a separate Python script, which can be called from a single main script. Below is the file structure we can use for a *Bokeh* application, adapted from the official documentation:

In order to generate a *Panel*-based web app, we have to make some changes in the code shown earlier. We create a file called ***scatterplot.py*** in the scripts folder. It is a modified and renamed ***main.py*** file presented in the section above, and it will contain the implementation code of the scatterplot. However, several changes in the code need to be done. One change is that we need to enclose the code into a newly defined function ***scatterplot_tab()*** to achieve some sort of encapsulation allowing us to handle it as a separate module. In order to visualize any displayable *Bokeh* model inside a *Panel*-based app, we need to make sure the *Bokeh* plot is in a *Panel pane* by calling *pn.pane.Bokeh(layout)*. The above steps can be repeated for all code corresponding to different plots or tabs. This approach ensures that we manage the modularized tabs as separate modules which can be selected in file *main.py*. Finally, we make the app executable on a *Bokeh* Server instance by annotating the objects to be deployed with a call to the *.servable()* method. The corresponding full source code can be inspected and downloaded from GitHub.

### 3.4 Adding a New Module in Panel-Based Web App

Functional analysis remains a critical step of the high-throughput studies. One essential resource for such analysis is Gene Ontology (GO), which provides a unified vocabulary to describe gene functions (GO terms). The authors performed a GO enrichment analysis over a set of DE genes to highlight groups of genes with coherent biological functions that are presumably acting in coordination in response to salt stress. Following the strategy described above (*see* Fig. 7), we intend to extend the functionality of the App. In order to illustrate how new modules can be easily added to the App, we provide another example by implementing a ***treemap***-module. The resulting *treemap*-tab represents visualization of a GO enrichment analysis findings published (*see* Fig. 8).

The resulting interactive dashboard comprising both plots can be accessed at the following binder link:

launch binder

**`https://mybinder.org/v2/gh/Alexyem1/bookchapter_example2/master?urlpath=%2Fproxy%2F5006%2Fpanel_app`**

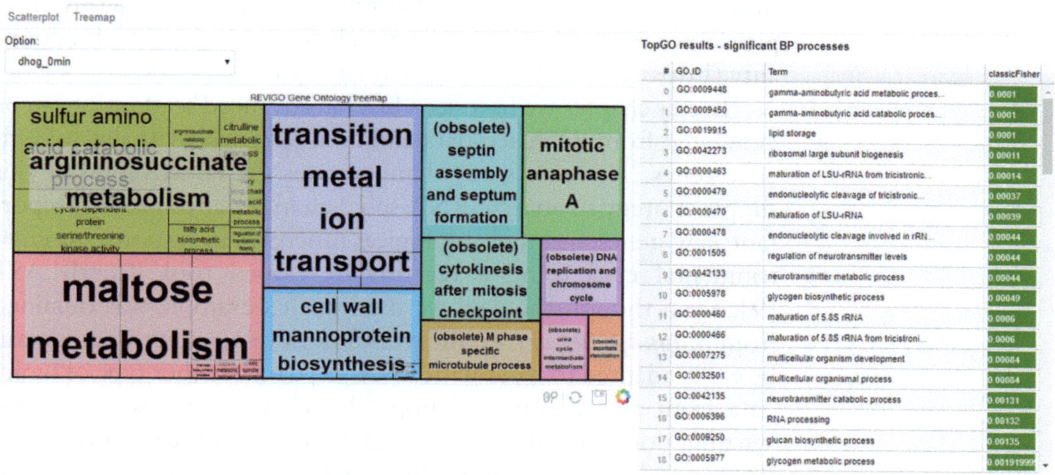


**Fig. 8** Visualization of treemap and associated dynamic table display GO enrichment analysis results for the strain *Δhog* upon salt stress after 25 min. The treemap visualization reflects results from REViGO analysis, whereas the table provides the underlying TopGO results. The most significant GO terms in the table based on the cut-off value of $>0.01$ are colored in green

The GO enrichment analysis was carried out for biological processes, utilizing Fisher's exact test with default parameters, with the help of R package topGO [10]. Finally, REViGO analysis was performed to reduce redundancy and visualize the overrepresented GO terms based on semantic similarity [11].

The *Bokeh*-App developed for this book chapter, the corresponding Jupyter Notebook complete with detailed comments, as well as the more advanced version of the *Panel*-based web app with implemented examples are available at https://github.com/Alexyem1/bookchapter. You can access these resources using Git (*see* Git website for instructions), or, alternatively, you can download the zip file hosted on GitHub.

### 3.5 Summary

The ability to explore data through quick and intuitive visualization is key for sophisticated data analysis (https://www.tapclicks.com/resources/blog/the-ultimate-guide-to-data-visualization/). There are many options for visualizing data within the Python ecosystem. The *Bokeh* library is a powerful tool for implementing visualizations that have a high degree of interactive capability, which can be customized even further using JavaScript [12]. Using *Bokeh*, it is possible to generate plots directly in Jupyter Notebook environment or create web applications for insightful visualizations. Similar to other frameworks like *Dash* (in particular *DashBio*) or *RStudio Shiny* package for the R environment, *Bokeh* includes dynamic filtering, selection, drop-downs, and other user interface elements, making for interactive visualizations that can be updated in real time. The interested reader may refer to the official website that provides useful tutorials and materials for exploring this functionality (https://docs.Bokeh.org/en/latest/).

In this chapter we have seen how an advanced scatterplot can be generated from RNA-seq data and used to quickly visualize significant genes. This case study demonstrates the effectiveness of this simple strategy to analyze RNA-seq data in order to predict important new genes and their biological functions in response to environmental cues using a recently developed tool PaperBLAST [9]. Even though the data we used in this example was obtained from *Magnaporthe oryzae*, the workflow developed in this guide can be easily adapted to work with RNA-seq data or other type of omics data from any organism. I hope this chapter represents a good starting point for the uninitiated to carry out data analysis of RNA-seq data. I also hope this article will be a useful guide for those interested in building their own custom visualizations to solve biological research problems.

## 4 Notes

1. The interested reader may explore the demo gallery to get a glimpse of how *Bokeh* works and to brainstorm some ideas for its potential. For further information, visit the official website (https://docs.Bokeh.org/en/latest/docs/gallery.html). Example projects are also available for experimentation on GitHub, each including a small dataset for testing.
2. The user should be aware that new program versions may be released in the future, resulting in different available options or software incompatibility.
3. The reader should be aware that the RNA-seq data presented here and derived from a previous study was based on DESeq2 calculation. The DESeq2 software utilized has parameters that can be optimized, and new tools for sequencing reads quantification appear regularly in the literature. The user should also be aware that changes in methods and parameters can impact the final results. A rigorous data analysis should entail evaluating multiple tools and options to identify the best performing pipeline according to the experimental design and questions.
4. The user should notice that data, methods, and parameters used at each step in this guide are intended as a simple first guideline for RNA-seq data analyses. More advanced features like sliders for considering of statistic values can be implemented to further extend the functionality of the web app presented.

5. Jupyter Notebook not only offers a simple way to visualize plots but also a convenient option for debugging code. To quickly reproduce and solve problems, it is recommended to develop charts in a Jupyter Notebook, as it can be frustrating to stop the server, make changes to the files, and restart the server to see whether the changes take effect. Jupyter Notebooks is an excellent environment for *Bokeh* development as it allows creating and testing fully interactive charts from a notebook. The syntax is a bit specific, but once you have a full plot, the code only needs to be changed slightly and can then be copied and pasted into a standalone *.py*-script, a strategy we presented and followed in this book chapter.
6. In order to produce the plot from this guide within the Jupyter Notebook environment, we need to define the corresponding output mode for our App. To do so, use function *output_notebook()* from *Bokeh.io* and *show(p)*, where *p* is our scatterplot. The corresponding Jupyter Notebook can be also found in the GitHub repository at https://github.com/Alexyem1/bookchapter.
7. *Note:* The plot will not display nor be interactive in the HTML version of the Jupyter Notebook; it will be interactive only as a standalone App. This is due to Python functions needed to be called to update the plot. If you desire interactivity in a notebook rendered into HTML, the interactions must be implemented using JavaScript. An alternative option is to use a Jupyter Notebook extension called *ipwidgets*. For more information and examples of using this tool, the interested reader may refer to a detailed documentation hosted at: https://ipywidgets.readthedocs.io/en/latest/.
8. The command *dirname(__file__)* allows to get an absolute path without relying on the working directory being the directory in which *main.py* of the app resides (→ this gives us more flexibility, for example, we can rename the app folder without affecting the functionality of the app). Thus, just uncomment the two lines of code in **step 2** and delete the preceding input data read-in commands used for the Jupyter Notebook environment. Furthermore, make sure to comment out *notebook_output()* and *show(p)* as we are no longer working in the Jupyter environment, thus making them redundant.

## Acknowledgments

This work was supported by IBWF gGmbH, UC Berkeley (Arkin Lab), and the BASF SE grant. In particular, I would like to sincerely thank Yanina Ageeva (IBM, Data Science and AI Elite) and Dr. Stefan Jacob (IBWF gGmbH) for critical reading of the manuscript.

## References


1. Dunn W Jr, Burgun A, Krebs MO, Rance B (2017) Exploring and visualizing multidimensional data in translational research platforms. Brief Bioinformatics 18(6):1044–1056
2. Komorowski M, Marshall DC, Salciccioli JD, Crutain Y (2016) Exploratory data analysis. In: Secondary analysis of electronic health records. Springer, Cham
3. Hossain S (2019) Visualization of bioinformatics data with dash bio:126–133. https://doi.org/10.25080/Majora-7ddc1dd1-012
4. Wang R, Perez-Riverol Y, Hermjakob H, Vizcaíno JA (2015) Open source libraries and frameworks for biological data visualisation: a guide for developers. Proteomics 15 (8):1356–1374
5. Antao T (2018) Bioinformatics with Python Cookbook: learn how to use modern Python bioinformatics libraries and applications to do cutting-edge research in computational biology. Packt Publishing Ltd, Birmingham
6. Hillebrand J, Nierhoff MH (2015) Mastering RStudio–develop, communicate, and collaborate with R. Packt Publishing Ltd, Birmingham
7. Pittard WS, Li S (2020) The essential toolbox of data science: Python, R, Git, and Docker. In: Computational methods and data analysis for metabolomics. Humana, New York, NY, pp 265–311
8. Bohnert S, Antelo L, Grünewald C, Yemelin A, Andresen K, Jacob S (2019) Rapid adaptation of signaling networks in the fungal pathogen *Magnaporthe oryzae*. BMC Genomics 20 (1):1–16
9. Price MN, Arkin AP (2017) PaperBLAST: text mining papers for information about homologs. MSystems 2(4):e00039-17
10. Alexa A, Rahnenfuhrer J (2019) Gene set enrichment analysis with topGO. Bioconductor Improv 27:1–12
11. Supek F, Bošnjak M, Škunca N, Šmuc T (2011) REVIGO summarizes and visualizes long lists of gene ontology terms. PLoS One 6(7): e21800
12. Embarak O (2018) Data visualization. In: Data analysis and visualization using python. Apress, Berkeley, CA

# Chapter 12

# Genome Editing Using CRISPR/Cas9 System in the Rice Blast Fungus

## Takayuki Arazoe

## Abstract

Fast and flexible genome manipulation is a powerful strategy for an in-depth understanding of molecular mechanisms in biological research. In recent years, CRISPR/Cas9-mediated genome editing has been used as a reliable genome manipulation method in a broad range of biological research including studies of filamentous fungi. The CRISPR/Cas9 system comprises a single-guide RNA (sgRNA) and a Cas9 protein, and the Cas9/sgRNA complex catalyzes a DNA double-strand break at the desired genomic locus. This protocol describes a fundamental CRISPR/Cas9 methodology that includes the design of the target sequence, construction of the CRISPR/Cas9 expression vector, and transformation for genome editing in *Pyricularia* (*Magnaporthe*) *oryzae*. This allows efficient targeted gene disruption, base editing, and reporter gene knock-in without any additional modifications of the host components. This protocol would be suitable for applying other CRISPR/Cas technologies and various functional genomics in *P. oryzae*.

**Key words** Genome editing, Functional genomics, CRISPR/Cas9, Homologous recombination, Filamentous fungi, *Pyricularia* (*Magnaporthe*) *oryzae*

## 1 Introduction

A flexible and efficient genome manipulation enables a deeper understanding of the fundamental molecular mechanisms in biological research. By emerging an RNA-guided engineered nuclease, clustered regularly interspaced short palindromic repeats (CRISPR)/CRISPR-associated protein (Cas) system [1], genome editing has become a promising approach for genome manipulation in a broad range of organisms and cell types. The CRISPR/Cas system is developed from a prokaryotic adaptive immune system and is categorized into two classes (classes 1 and 2) and six types (types I–VI) [2, 3]. The class 2/type II CRISPR/Cas9 system, derived from *Streptococcus pyogenes*, is the most characterized and widely used as the genome editing tool in various kinds of research [4, 5] since this system comprises only two components, a Cas9 and

Stefan Jacob (ed.), *Magnaporthe oryzae: Methods and Protocols*, Methods in Molecular Biology, vol. 2356,
https://doi.org/10.1007/978-1-0716-1613-0_12, 

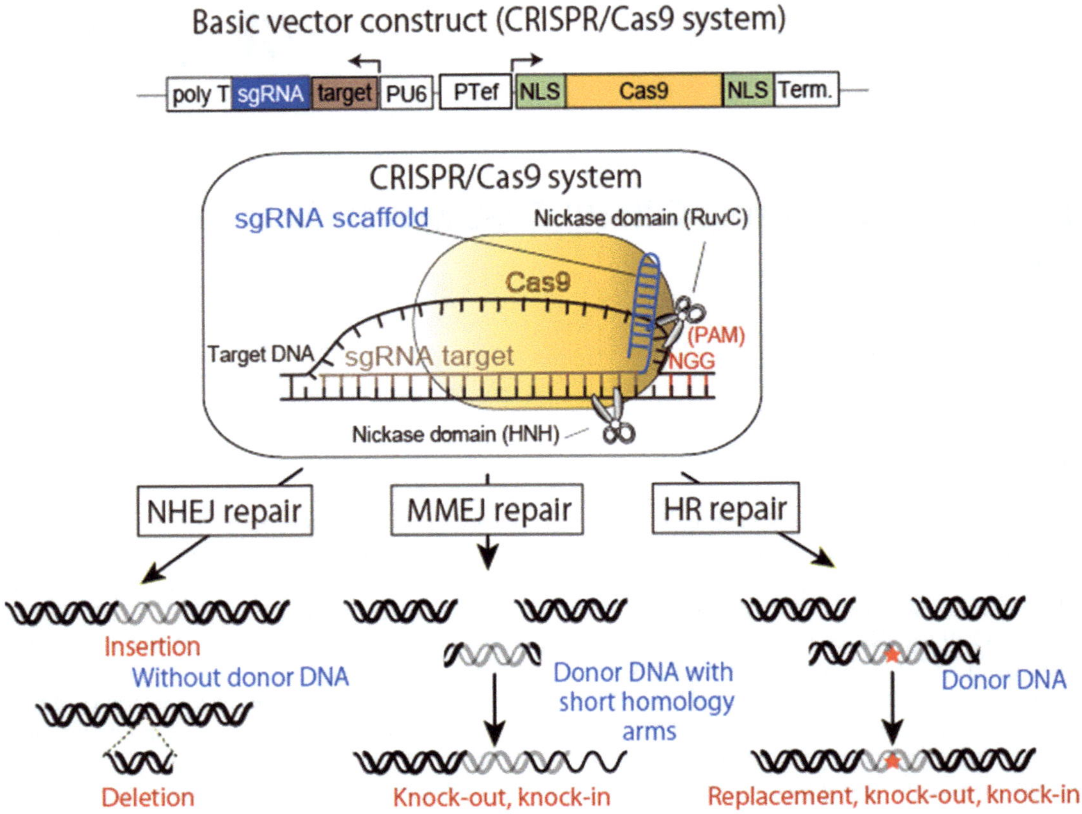


**Fig. 1** Schematic representation of DNA double-strand break-mediated genome editing using CRISPR/Cas9 system in filamentous fungi. The basic construct of the CRISPR/Cas9 expression vector is depicted here (*see* **Note 1**). The CRISPR/Cas9 expression vector has two components: single-guide RNA (sgRNA) and Cas9 expression cassette. For effective expression of sgRNA in the nucleus, an endogenous small nuclear U6 RNA gene promoter (U6 promoter), which is an RNA polymerase III-dependent promoter (*see* **Note 1**), can be used. For the strong expression of Cas9, Tef (translation elongation factor) promoter, which is a constitutive expression promoter (*see* **Note 1**), can be used. The nuclear localization signal (NLS) derived from simian virus 40 is added to the N-terminus and C-terminus of the Cas9. PTef indicates the Tef promoter. PU6 indicates the endogenous U6 promoter. Poly T indicates the sgRNA terminator (5′-TTTTT-3′). The term indicates glucoamylase terminator. The DNA double-strand break (DSB) induced by the sgRNA/Cas9 complex is illustrated on the bottom of the vector scheme. The DSB is largely repaired by non-homologous end joining (NHEJ), microhomology-mediated end joining (MMEJ), or homologous recombination (HR). These repair pathways allow induction of various types of mutations at the target genomic locus. PAM indicates protospacer adjacent motif

a single-guide RNA (sgRNA). The Cas9/sgRNA complex requires the 5′-NGG-3′ motif, known as the protospacer adjacent motif (PAM), and catalyzes a DNA double-strand break (DSB) by forming an sgRNA/target DNA hybrid (20 bp) upstream of the complementary PAM sequence (Fig. 1). Therefore, a programmable nuclease targeting the desired genomic locus can be constructed by one-step replacement of the target sequence of sgRNA. In

*P. oryzae*, CRISPR/Cas9-mediated genome editing has been performed by using the optimized all-in-one vector system [6] and ribonucleoprotein method [7].

Fundamental genome editing involves triggering the host DNA repair by cleaving the targeted genomic locus (Fig. 1).

Inducing non-homologous end joining (NHEJ), homologous recombination (HR), and microhomology-mediated end joining (MMEJ) repairs via CRISPR/Cas9 has been applied for fungal genome editing, including marker-free and multiplex gene disruption (Fig. 1) [8–10]. The rates of these repair pathways depend on each host genome property. In plants and mammalian cells, the NHEJ-mediated gene disruption has been used as a simple and efficient genome editing strategy by integrating the CRISPR/Cas9 vector into the genome. Meanwhile, in several filamentous fungi, the excessive DSBs without the donor DNA templates often lead to cell toxicity (severe off-target effects and escape transformants) and unexpected chromosomal large deletion [7, 11–13], which resulted in the inaccurate and less efficient genome editing. However, in most filamentous fungi, HR-mediated genome editing can be performed with higher efficiency than that of the other eukaryotes. HR-mediated genome editing enables to introduce the desired mutations at the target locus, double crossover recombination (gene conversion) enables targeted gene disruption (knock-out), and single crossover type recombination enables perfectly designed base editing, reporter gene knock-in, and gene disruption by the simple donor vector construct (Fig. 2) [6, 14].

HR can be induced via transient expression of the CRISPR/Cas9 system in the nucleus of protoplasts, which enables to elevate the efficiency of genome-edited transformants and to minimize the off-target effects (Fig. 3).

Based on these findings, this chapter describes the standard protocol for HR-mediated genome editing using the CRISPR/Cas9 system in *P. oryzae*, which can be applied for a wide range of research and expand other genome editing technologies. The method presented includes the design of the target sequence, construction of the CRISPR/Cas9 expression vector, and transformation of *P. oryzae*.

## 2 Materials

High-grade chemicals are the preferred choice. General lab equipment such as pipettes, scales, stir sticks, thermocycler, heat block, incubator, vortex mixer, centrifuge, and clean bench are not mentioned.

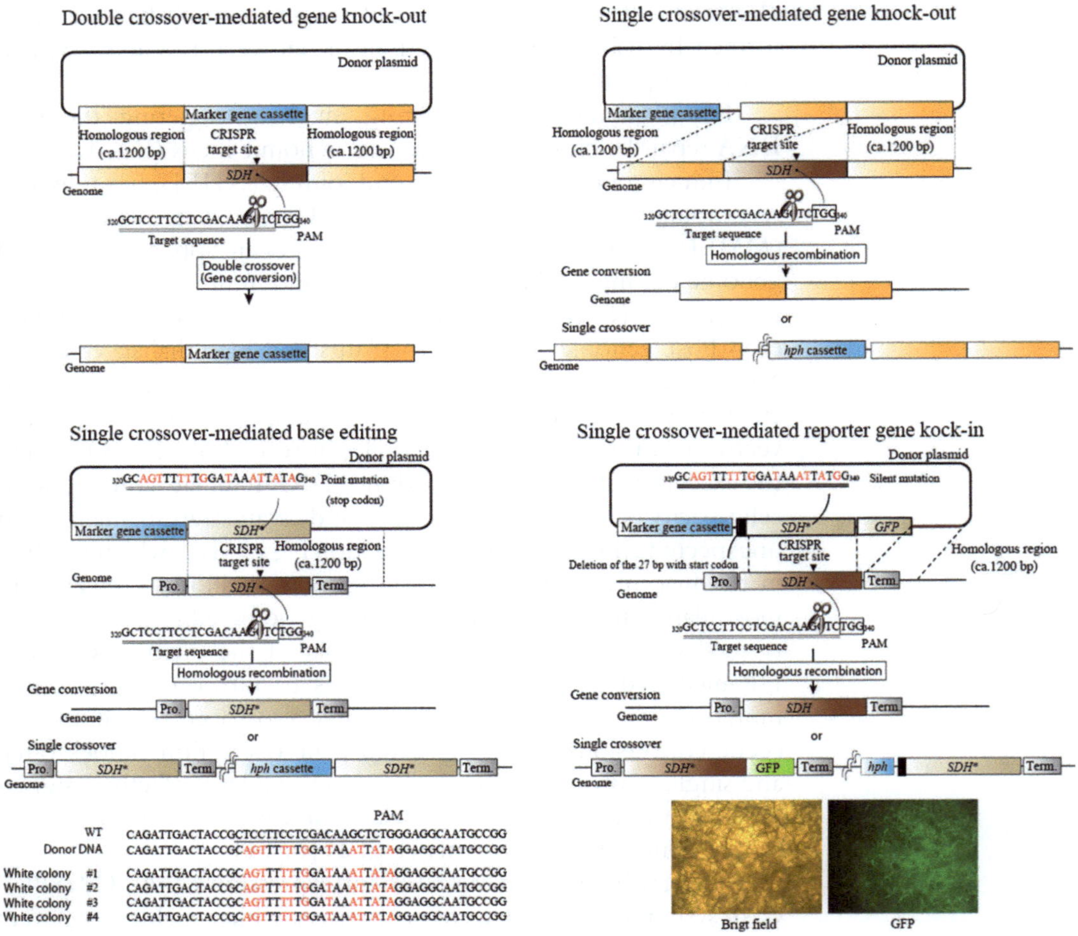


**Fig. 2** Genome editing via CRISPR/Cas9-mediated double crossover (gene conversion) and single crossover type recombination. Double crossover type recombination can be applied for precise and efficient gene disruption via CRISPR/Cas9. Simultaneous introduction of four vectors, two CRISPR/Cas9 vectors and two donor vectors, can be applied for the one-step double-gene disruption. In addition, CRISPR/Cas9-mediated single crossover type recombination can also be applied for targeted base editing, reporter gene knock-in, and gene disruption. Since crossover occurs after repair of the cleaved DNA, recombination products contain the designed mutations, and the whole vector sequence is integrated into the desired locus. The donor vectors can be easily constructed by standard strategy, such as fusion PCR, In-Fusion, or Gibson assembly-based cloning methods

### *2.1 Design of Target Sequence of sgRNA*

1. Search engine (CRISPR direct) (https://crispr.dbcls.jp).
2. Target DNA (gene) sequence.

### *2.2 Cloning Designed Target Sequence of sgRNA into Backbone Vector*

1. Two designed oligos for cloning the designed target sequence of sgRNA.
2. All-in-one CRISPR/Cas9 backbone vector (*see* **Note 1**).
3. Esp3I (BsmBI) (Thermo Fisher Scientific) (strongly recommend).

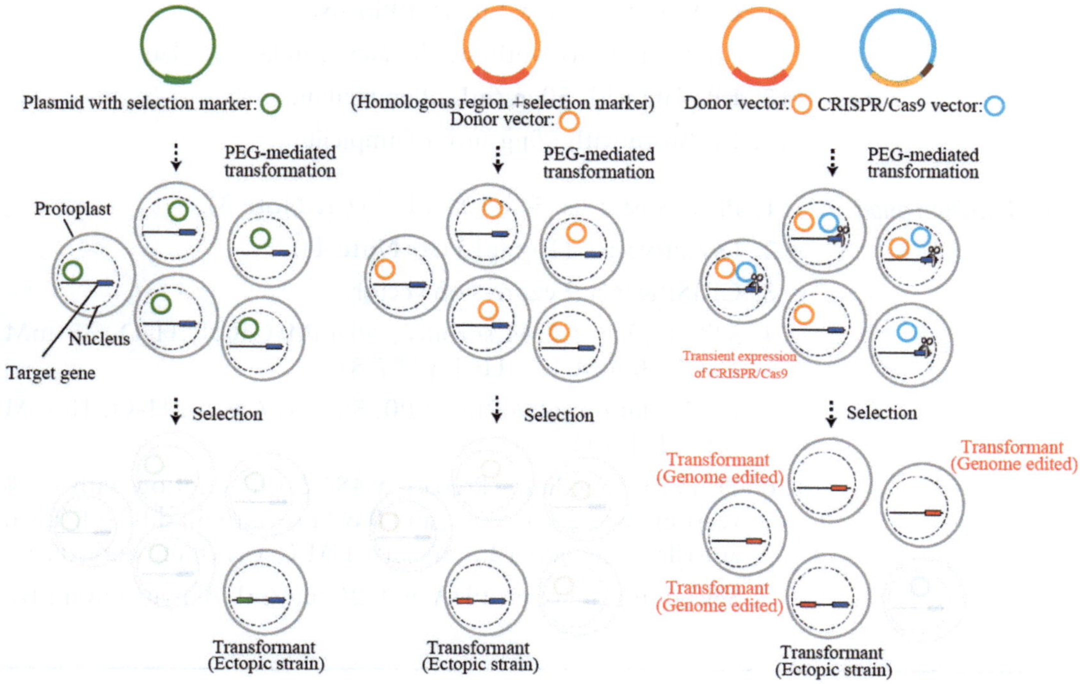


**Fig. 3** HR-mediated genome editing by transient expression of CRISPR/Cas9 system. By the polyethylene glycol (PEG)-mediated transformation, exogenous plasmids (DNAs) can be introduced into the cells, and then parts of the plasmids can also be introduced into the nucleus (left). The plasmid vectors existing in the nucleus are randomly integrated into the cleaved genomic loci via NHEJ repair, which resulted in the transformants (ectopic strain) (left). Use of the vector harboring homologous regions of the target gene (donor vector) can induce HR between the donor vector and target gene locus, which resulted in the target gene disruption. However, the HR in the *P. oryzae* genome tends to be at low efficiency without the induction of the site-specific DSB (center). Simultaneous introduction of the CRISPR/Cas9 vector and donor vector can induce the efficient target gene cleavage and HR between the donor vector and target gene locus (right). In addition, transient expression of CRISPR/Cas9 vector existing in the nucleus induces site-specific DSB, which can increase the efficiency of the target gene disruptions and the number of transformants (right). Transient expression of the nuclease would be lower off-target effects than that of constitutive expression by the integration of the nuclease gene into the genome

4. Quick Ligation Kit (New England Biolabs) (strongly recommend).
5. T4 DNA Ligation Buffer (New England Biolabs) (strongly recommend).
6. T4 polynucleotide kinase (New England Biolabs) (strongly recommend).
7. *E. coli* competent cells.
8. DNA polymerase (for colony PCR).
9. PCR primer (5′- ATCTGAACCATCCTTTGAC CACCGTTTGCT -3′) (for colony PCR) and the designed oligo bottom (*see* **Note 2**).
10. Plasmid Midiprep kit.

11. Cas9 protein (New England Biolabs).
12. sgRNA in vitro synthesis kit (New England Biolabs).
13. LB plate with 50μg/mL of ampicillin.
14. LB Broth with 50μg/mL of ampicillin.

### 2.3 Transformation

1. Protoplast cells ($5 \times 10^5$ $mL^{-1}$) (*see* **Note 3**).
2. Donor vector (1μg/μL) (*see* **Note 4**).
3. CRISPR/Cas9 expression vector.
4. STC buffer. 1.2 M sorbitol, 50 mM $CaCl_2 \cdot 2H_2O$, 35 mM NaCl, 10 mM Tris–HCl (pH 7.5).
5. PEG solution. 60% PEG4000, 50 mM $CaCl_2 \cdot 2H_2O$, 10 mM Tris–HCl (pH 7.5).
6. YG1/2SC medium. Prepare a 487.5 mL solution with 0.5% yeast extract, 2% glucose, and 0.6 M sorbitol in $ddH_2O$. Then autoclave and add 12.5 mL of a 1 M $CaCl_2$ (filter sterilized).
7. Selection plate (e.g., PDA with 200μg/mL of hygromycin B).

## 3 Methods

HR-mediated genome editing using CRISPR/Cas9 system in *P. oryzae* involves a four-step procedure: (1) design of target sequence of sgRNA, (2) cloning of the designed target sequence into the backbone vector, (3) transformation, and (4) checking the genome-edited strain. In the case of genome editing using Cas9 protein and transcribed sgRNA, the procedure in **step 2** is replaced with "in vitro sgRNA synthesis." In **step 2**, the type IIS restriction enzyme (RE) Esp3I is used for simultaneous reaction of RE digestion and ligation, which is applied for the construction of the CRISPR/Cas9 expression vector library. If the final sgRNA sequence is identical, following the procedure in **step 2** is not required. As **step 4** can be performed using standard PCR and/or sequencing analysis, a detailed protocol is not described in this chapter. Here, as an example, *scytalone dehydratase* (*SDH*) gene modification methods are described.

### 3.1 Design of Target Sequence of sgRNA

1. Start the search engine (CRISPR direct) (https://crispr.dbcls.jp).
2. Paste a target DNA sequence or upload a sequence file. The accession number or genome location of NCBI is also used to design the sequence.
3. Choose "Rice blast fungus (*Magnaporthe oryzae*) 70–15 genome, MG8 (Sep, 2011)" as the "Specificity check." "PAM sequence requirement" is NGG (default).
4. Click the "design" button.

Show 20 entries

Search:

| position | | target sequence | sequence information | | | | number of target sites | | |
|---|---|---|---|---|---|---|---|---|---|
| start - end | + − | 20mer+PAM (total 23mer) | GC% of 20mer | Tm of 20mer | TTTT in 20mer | restriction sites | 20mer +PAM | 12mer +PAM | 8mer +PAM |
| 135 - 157 | - | CCT CATGACTTGCGTCTATGAGT [gRNA] | 45.00 % | 68.80 °C | - | BspHI | 1 [detail] | 1 [detail] | 46 [detail] |
| 137 - 159 | + | TCATGACTTGCGTCTATGAG TGG [gRNA] | 45.00 % | 68.35 °C | - | BspHI | 1 [detail] | 1 [detail] | 29 [detail] |
| 138 - 160 | + | CATGACTTGCGTCTATGAGT GGG [gRNA] | 45.00 % | 68.80 °C | - | | 1 [detail] | 1 [detail] | 25 [detail] |
| 159 - 181 | + | GGCAGACAGCTACGACTCCA AGG [gRNA] | 60.00 % | 77.35 °C | - | | 1 [detail] | 1 [detail] | 153 [detail] |
| 164 - 186 | + | ACAGCTACGACTCCAAGGAC TGG [gRNA] | 55.00 % | 74.98 °C | - | | 1 [detail] | 3 [detail] | 144 [detail] |
| 165 - 187 | + | CAGCTACGACTCCAAGGACT GGG [gRNA] | 55.00 % | 74.89 °C | - | | 1 [detail] | 2 [detail] | 185 [detail] |
| 170 - 192 | + | ACGACTCCAAGGACTGGGAT AGG [gRNA] | 55.00 % | 76.03 °C | - | | 1 [detail] | 1 [detail] | 77 [detail] |
| 176 - 198 | - | CCA AGGACTGGGATAGGCTGCGA [gRNA] | 60.00 % | 79.46 °C | - | | 1 [detail] | 1 [detail] | 121 [detail] |
| 180 - 202 | + | GGACTGGGATAGGCTGCGAA AGG [gRNA] | 60.00 % | 78.16 °C | - | | 1 [detail] | 1 [detail] | 100 [detail] |
| 211 - 233 | - | CCT ACTCTGCGCGTATGTTCCGC [gRNA] | 60.00 % | 77.42 °C | - | | 1 [detail] | 1 [detail] | 64 [detail] |
| 230 - 252 | - | CCG CCCTGCCATGTTTATTTTTA [gRNA] | 35.00 % | 65.26 °C | - | | 1 [detail] | 1 [detail] | 126 [detail] |
| 233 - 255 | - | CCC TGCCATGTTTATTTTTACTT [gRNA] | 25.00 % | 60.12 °C | - | | 1 [detail] | 1 [detail] | 103 [detail] |
| 234 - 256 | - | CCT GCCATGTTTATTTTTACTTT [gRNA] | 25.00 % | 58.75 °C | - | | 1 [detail] | 1 [detail] | 121 [detail] |
| 238 - 260 | - | CCA TGTTTATTTTTACTTTCCCA [gRNA] | 25.00 % | 59.71 °C | - | | 1 [detail] | 2 [detail] | 102 [detail] |
| 257 - 279 | - | CCC ACACCAAATCCAGACTTTAA [gRNA] | 35.00 % | 65.60 °C | - | | 1 [detail] | 1 [detail] | 174 [detail] |
| 258 - 280 | - | CCA CACCAAATCCAGACTTTAAC [gRNA] | 40.00 % | 65.60 °C | - | | 1 [detail] | 2 [detail] | 89 [detail] |
| 263 - 285 | - | CCA AATCCAGACTTTAACAGCGA [gRNA] | 40.00 % | 67.90 °C | - | | 1 [detail] | 2 [detail] | 82 [detail] |
| 269 - 291 | - | CCA GACTTTAACAGCGACGACCA [gRNA] | 50.00 % | 71.39 °C | - | | 1 [detail] | 1 [detail] | 38 [detail] |
| 289 - 311 | - | CCA AAAAAAAAAAAAAAAACAGA [gRNA] | 10.00 % | 47.78 °C | + | | 2 [detail] | 526 [detail] | 1716 [detail] |
| 319 - 341 | - | CCG CTCCTTCCTCGACAAGCTCT [gRNA] | 55.00 % | 74.29 °C | - | | 1 [detail] | 2 [detail] | 195 [detail] |

Showing 21 to 40 of 105 entries

First Previous 1 2 3 4 5 Next Last

**Fig. 4** Search result of target sequence by using CRISPR direct. In this case, the search results of the target sequence of CRISPR/Cas9 for the *SDH* gene (from start codon to stop codon) are shown. "Start-end" indicates the target sequence region in the input sequence. "+ −" indicates that the target sequence with PAM exists on the plus (+) or minus (−) strand of the input sequence. "GC% of 20mer" indicates GC% of the target sequence. "Tm of 20mer" indicates Tm value of the target sequence. "TTTT in 20mer" indicates that TTTT sequence (stop the transcription) exists in the target sequence. "restriction sites" indicates the RE enzyme name existing in the target sequence. "20mer + PAM" indicates the number of the site of the target sequence with PAM in the whole genome sequence of *P. oryzae.* "23mer + PAM" indicates the number of the site of the target sequence (12-mer target sequence flanking with PAM) with PAM. "8mer + PAM" indicates the number of sites of the target sequence (8-mer target sequence flanking with PAM) with PAM. The target sequences that are highly specific and have fewer off-target sites are highlighted. By clicking the "detail" of the "number of target sites," the detailed sequence of the target site can be shown

5. Select the target sequence with the lowest number of "20mer + PAM," "12mer + PAM," and "8mer + PAM" (Fig. 4) (*see* **Note 5**).
6. Design and order the oligos (designed oligo top, 5′-ttcgCAT GACTTGCGTCTATGAGT -3′; designed oligo bottom, 5′-aaac ACTCATAGACGCAAGTCATG -3′) for cloning the target sequence of sgRNA into the CRISPR/Cas9 backbone vector (Fig. 5).

***3.2 Cloning of Designed Oligos into the Backbone Vector***

1. In a PCR tube, mix the ingredients in the following order for the two designed oligos annealing: 2μL designed oligo top (50μM), 2μL of designed oligo bottom (50μM), 1μL of T4 ligase buffer (10×), 1μL of T4 polynucleotide kinase, and 4μL $ddH_2O$.

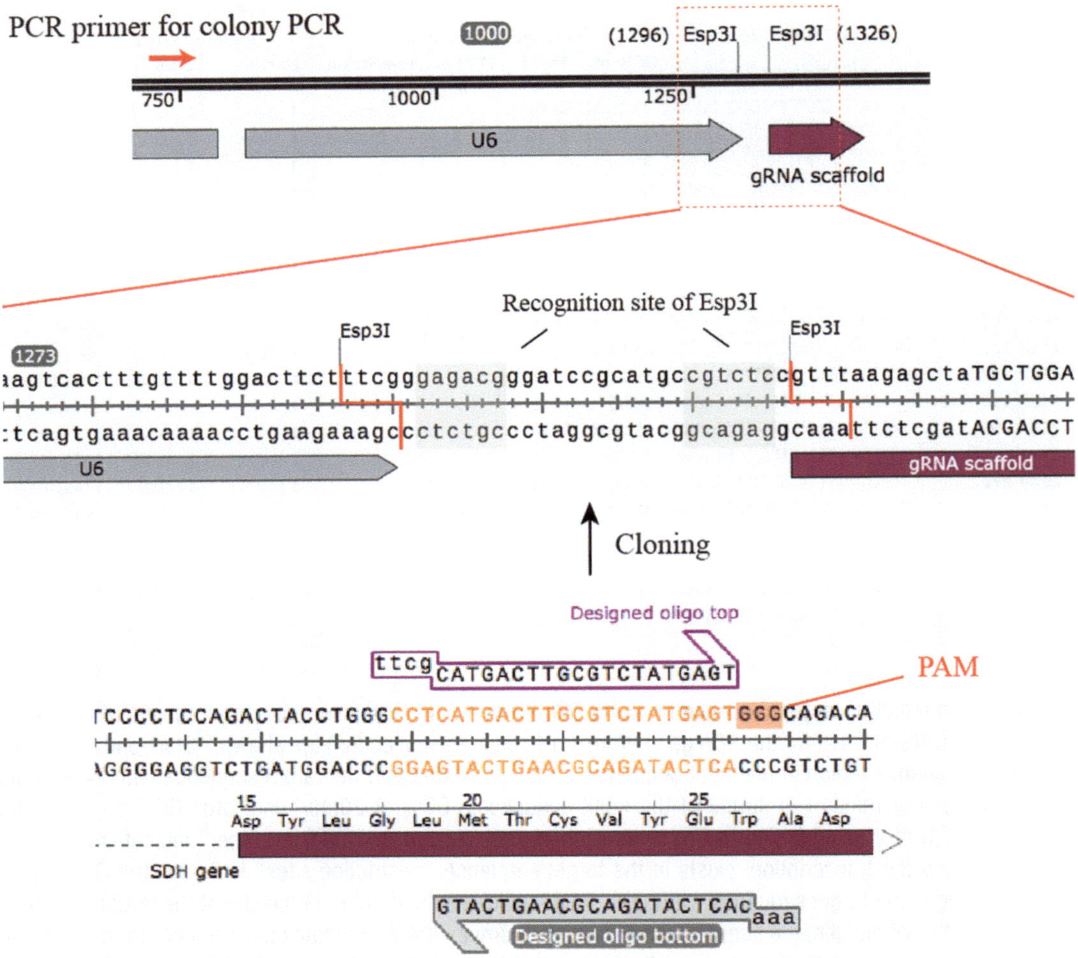


**Fig. 5** Schematic representation of design and cloning of target sequence. Schematic representation of the U6 promoter, cloning site (Esp3I), sgRNA scaffold, and PCR primer position for colony PCR (upper panel). Schematic representation of the target sequence and designed oligos for the *SDH* gene cleavage (lower panel). By the digestion of the CRISPR/Cas9 backbone vector with Esp3I, two sticky ends (U6 side, 5′-CGAA-3′; sgRNA scaffold side, 5′GTTT-3′) appear in the backbone vector. By adding the complimentary sequence of the sticky ends of the backbone vector to oligos of the target sequence, the annealed-oligo can be cloned into the Esp3I sites of the backbone vector. After cloning, the positive colony can be detected by PCR using the PCR primer and designed oligo bottom

2. Set the PCR tube in the thermocycler and program as follows: 37 °C for 30 min, 95 °C for 5 min, ramp down to 25 °C at 5 °C/min.
3. Dilute the annealed-oligo solution in the ratio 1:20 by adding 1μL of the annealed-oligo solution to 19μL dd$H_2O$.
4. In a PCR tube, mix the ingredients in the following order (*see* **Note 6**): 0.8μL of the diluted annealed-oligo, 0.3μL of

CRISPR/Cas9 backbone vector (50 ng/μL), 0.5μL of T4 ligase buffer (10×), 0.2μL of Esp3I, 0.2μL of Quick Ligase, and 3.1μL of $ddH_2O$.

5. Set the PCR tube in the thermocycler and program as follows: (37 °C for 5 min = > 16 °C for 10 min) × 6 cycle, 37 °C for 30 min.
6. Add 2μL of the digest reaction to 20μL of *E. coli* competent cells, and then follow the standard heat shock transformation protocol. Spread the *E. coli* cells onto an amp selection LB plate and incubate overnight at 37 °C.
7. After incubation, pick 4–10 colonies (or more) on LB plate with 50μg/mL of ampicillin and perform colony PCR using the specific primer (5′- ATCTGAACCATCCTTTGAC CACCGTTTGCT -3′) and the designed oligo bottom (5′-aaac NNNNNNNNNNNNNNNNNNNN -3′). In this case (targeting *SDH*), the designed oligo bottom (5′-aaacACT CATAGACGCAAGTCATG-3′) can be used as a PCR primer. The positive band shows about 600 bp (*see* **Note 7**).
8. Incubate *E. coli* cells, which showed a positive signal by colony PCR, in LB Broth with 50μg/mL of ampicillin and isolate the plasmid using Midiprep kit following the protocol (*see* **Note 8**).

### 3.3 Transformation of P. oryzae

Carry out all the steps that follow inside a clean bench. The protoplasts and donor vector should be prepared following the standard protocol.

1. Thaw tubes containing the protoplast solution of *P. oryzae* on ice.
2. Add 2.5μL of the CRISPR/Cas9 vector (1μg/μL) and 2.5μL of the donor vector (1μg/μL) to a 1.5 mL tube, and gently mix by pipetting. In the case of the use of Cas9 protein and in vitro transcribed sgRNA, add 1.5μg of the Cas9 protein and 300 ng of the sgRNA to a 1.5 mL tube, and gently mix by pipetting. Incubate at room temperature for 15 min. After incubation, add 2.5μL of the donor vector (1μg/μL) to the ribonucleoprotein mixture.
3. Add 50μL of the protoplast solution ($5 \times 10^5$/mL) to the mixture, and gently mix by pipetting.
4. Add 50μL of PEG solution to the mixture. Mix the PEG solution and the mixture thoroughly by flicking the tube and pipetting gently.
5. Place the mixture at room temperature for 20 min.
6. Add 1 mL of STC solution and mix by flicking and inverting the tube.
7. Centrifuge the tube at 4000 × *g* for 5 min at 4 °C.

8. Discard the supernatant by decantation. Add YG1/2SC medium and mix well by pipetting gently.
9. Incubate the tube at 28 °C for 12–24 h (*see* **Note 8**).
10. Plate the regenerated cells onto the optimal selection plate (*see* **Note 9**).
11. After 4–7 days post-plating, the resistant colonies appear. Check the accurate genome editing had occurred by PCR and sequencing analysis.

## 4 Notes

1. We can provide the CRISPR/Cas9 backbone vector which is fully optimized for *P. oryzae*. Improved CRISPR/Cas9 systems, such as modified sgRNA scaffold [15], modified NLS, and modified Cas9 (ex. high-fidelity) [16], and other types of CRISPR/Cas systems (ex. Cpf1) can also be provided for increasing the specificity, decreasing the off-target effects, and resolving the escape mechanism. The endogenous U6 promoter for sgRNA expression can be obtained from the genome of *P. oryzae*. The Tef promoter can be replaced with the other strong constitutive expression promoters. The DNA sequence of the basic vector including the promoter and terminator can be obtained from supplementary information of reference [6].
2. The one of two designed oligos containing the sgRNA scaffold side of the sticky end (5′-aaac-3′) can also be used as the primer for colony PCR. PCR primer (5′- ATCTGAACCATCCTTT GACCACCGTTTGCT-3′) anneals to the top of the Tef promoter sequence (Fig. 5).
3. The protoplasts obtained from the standard protocols can be used for genome editing of *P. oryzae*.
4. The two types of the donor vector can be used for genome editing: double crossover (gene conversion) type and single crossover type vectors (Fig. 2), which can be constructed using standard cloning methods, such as RE-based, In-Fusion, and Gibson assembly methods. It is recommended that the homologous sequence should be longer than 1000 bp, though a shorter homologous sequence can be used for genome editing. To design the donor vector for single crossover-mediated base editing, at least three mutations including the desired mutation should be introduced into the homologous sequence corresponding with the CRISPR/Cas9 target site (about 10 bp from PAM) to avoid the cleavage. When the CRISPR target site cannot be designed at the mutation site, silent mutations should be introduced into the target site of homologous sequence. Besides, designed mutations should be introduced

into the within around of the 100 bp at the CRISPR target site. To design the donor vector for single crossover-mediated reporter gene knock-in, the promoter region and start codon of homologous sequence would be removed to avoid the duplication of the target gene cassette, and the stop codon should also be removed from homologous sequence to fuse to the reporter gene.

5. Since the genome DNA sequence of the 70–15 strain is used as a reference sequence, the "number of target sites" may indicate "0" when the DNA sequence of the other *P. oryzae* strain was inputted. In this case, "0" indicates the lowest number of the off-target candidates. It is preferable to avoid the target sequence containing extremely high or low GC% and the Tm value and "TTTT" sequence, which would stop the transcription.
6. This step includes simultaneous *Esp*3I digestion and DNA ligation reaction. Since Esp3I recognizes asymmetric DNA sequence and cleaves outside of the recognition sequence, the recognition sites are removed from the ligated products.
7. If the positive band is not obtained by colony PCR, scale up all the reaction orders and retry.
8. If the sequencing analysis is performed, the specific primer (5′-ATCTGAACCATCCTTTGACCACCGTTTGCT-3′) can also be used for this analysis.
9. The protoplast solution can be regenerated and selected by using the high osmotic medium.

## Acknowledgments

This work is supported by the Japan Society for the Promotion of Science (JSPS) Grant-in-Aid for Scientific Research (A) (Grant Number 16H02536) and Grant-in-Aid for Young Scientists (B) (Grant Number 15 K18647).

## References

1. Gasiunas G, Barrabgou R, Horvath P, Siksnys V (2012) Cas9-crRNA ribonucleoprotein complex mediates specific DNA cleavage for adaptive immunity in bacteria. Proc Natl Acad Sci U S A 109:15539–15540
2. Makarova KS, Zhang F, Koonin EV (2017) SpapShot: class 1 CRISPR-Cas systems. Cell 168:946–946
3. Makarova KS, Zhang F, Koonin EV (2017) SpapShot: class 2 CRISPR-Cas systems. Cell 168:328–328
4. Jinek M, Chylinski K, Fonfara I, Hauer M, Doudna JA, Charpentier E (2012) A programmable dual-RNA-guided DNA endonuclease in adaptive bacterial immunity. Science 337:816–821
5. Adli M (2018) The CRISPR tool kit for genome editing and beyond. Nat Commun 9:1911
6. Arazoe T, Miyoshi K, Yamato T, Ogawa T, Ohsato S, Arie T, Kuwata S (2015) Tailor-made CRISPR/Cas system for highly efficient

targeted gene replacement in the rice blast fungus. Biotechnol Bioeng 112:2543–2549

7. Foster AJ, Martin-Urdiroz M, Yan X, Wright HS, Soanes DM, Talbot NJ (2018) CRISPR-Cas9 ribonucleoprotein-mediated co-editing and counterselection in the rice blast fungus. Sci Rep 8:14355
8. Wakai S, Arazoe T, Ogino C, Kondo A (2017) Future insights in fungal metabolic engineering. Bioresour Technol 245:1314–1326
9. Song R, Zhai Q, Sun L, Huang E, Zhang Y, Zhu Y, Guo Q, Tian Y, Zhao B, Lu H (2019) CRISPR/Cas9 genome editing technology in filamentous fungi: progress and perspective. Appl Microbiol Biotechnol 103:6919–6932
10. Schuster M, Kahmann R (2019) CRISPR-Cas9 genome editing approaches in filamentous fungi and oomycetes. Fungal Genet Biol 130:43–53
11. Arazoe T, Ogawa T, Miyoshi K, Yamato T, Ohsato S, Sakuma T, Yamato T, Arie T, Kuwata S (2015) Tailor-made TALEN system for highly efficient targeted gene replacement in the rice blast fungus. Biotechnol Bioeng 112:1335–1342
12. Mizutani O, Arazoe T, Toshida K, Hayashi R, Ohsato S, Sakuma T, Yamamoto T, Kuwata S, Yamada O (2017) Detailed analysis of targeted gene mutations caused by the platinum-fungal TALENs in *Aspergillus oryzae* RIB40 strain and a ligD disruptant. J Biosci Bioeng 123:287–293
13. Zhang C, Meng X, Wei X, Lu L (2016) Highly efficient CRISPR mutagenesis by microhomology-mediated end joining in *Aspergillus fumigatus*. Fungal Genet Biol 86:47–57
14. Yamato T, Handa A, Arazoe T, Kuroki M, Nozaka A, Kamakura T, Ohsato S, Arie T, Kuwata S (2019) Single crossover-mediated targeted nucleotide substitution and knock-in strategies with CRISPR/Cas9 system in the rice blast fungus. Sci Rep 9:7427
15. Chen B, Gilbert LA, Cimini BA, Schnitzbauer J, Zhang W, Li GW, Park J, Blackburn EH, Weissman JS, Qi LS, Huang B (2014) Dynamic imaging of genomic loci in living human cells by an optimized CRISPR/Cas system. Cell 156:373
16. Vakulskas CA, Dever DP, Rettig GR, Turk R, Jacobi AM, Collingwood MA, Bode NM, McNeill MS, Yan S, Camarena J, Lee CM, Park SH, Wiebking V, Bak RO, Gomez-Ospina N, Pavel-Dinu M, Sun W, Bao G, Porteus MH, Behlke MA (2018) A high-fidelity Cas9 mutant delivered as a ribonucleoprotein complex enables efficient gene editing in human hematopoietic stem and progenitor cells. Nat Med 24:1216–1224

# Chapter 13

# Diurnal, Circadian, and Photomorphogenic Analyses in *Magnaporthe oryzae*


## Ciarán Griffin and George R. Littlejohn



## Abstract

Circadian rhythms have been shown to play an important role in plant-pathogen interactions between plants and fungi. These protocols describe the methodology used to determine the function and characteristics of the diurnal and circadian behavior in the hemibiotrophic fungal pathogen, *Magnaporthe oryzae*. Here we describe methods to determine how conidiation, conidial development, and pathogenicity may be altered in *M. oryzae* as a result of differing diurnal or circadian environmental conditions.



**Key words** Circadian, Diurnal, Time of day, Clock input/output, Conidiation, Conidial development, Pathogenicity


## 1 Introduction

As the Earth rotates on its axis once every 24 h, living organisms experience a day-night cycle and have evolved mechanisms to coordinate aspects of their growth, development, and behavior to their rhythmically fluctuating surroundings. Environmental inputs such as light quality and intensity, temperature, and humidity fluctuate on a daily basis, and terrestrial organisms alter their metabolism, physiology, and behavior to perform optimally under changing conditions [1–3]. These oscillating daily cues give rise to diurnal (daily (L.)) rhythms which persist only in the presence of rhythmic environmental cues [4]. In addition to responding to diurnal rhythms directly, many organisms have developed the ability to measure time based on external daily cues, anticipate the daily 24-h period, and adjust these aspects of their physiology and cellular function accordingly [5, 6]. An internal anticipatory oscillation is known as a circadian rhythm, derived from the Latin "circa"

**Electronic supplementary material:** The online version of this chapter (https://doi.org/10.1007/978-1-0716-1613-0_13) contains supplementary material, which is available to authorized users.


Stefan Jacob (ed.), *Magnaporthe oryzae: Methods and Protocols*, Methods in Molecular Biology, vol. 2356,
https://doi.org/10.1007/978-1-0716-1613-0_13, © Springer Science+Business Media, LLC, part of Springer Nature 2021

(about) and "dies" (day) [7]. Circadian rhythms are established, or entrained, by reliable rhythmic environmental phenomena such as daily fluctuations in light intensity or temperature. Circadian rhythms are generated in cells endogenously and can be maintained under constant, free-running conditions, where there are no external, diurnal, or temporal cues such as light or temperature to input, reset, and entrain the clock [4, 7–11].

Circadian rhythms then confer on biological organisms the ability to reliably predict future environmental conditions and thereby optimize metabolic function, physiology, and behavior to anticipate predictable future challenges. Circadian clock outputs include rhythmic control of response to abiotic stress and immune response in plants [11–15] and, additionally, contribute to pathogenicity and virulence in fungal pathogens [11, 14, 16], such as *Botrytis cinerea.*

Most described circadian oscillators rely on a network of integrated transcriptional/translational feedback loops. At its core, there is generally a positive and negative arm of the network [2, 17]; here, the transcription factor-based positive arm initiates the transcription of the negative arm, which, when translated, suppresses the action of the positive arm, repressing its own transcription [2]. Throughout the circadian cycle, the negative arm proteins are modified, often through phosphorylation or ubiquitination [15, 18], facilitating protein turnover in the proteasome [2]. Consequent relief of repression of the positive arm leads to the cycle to beginning anew after approximately 24 h.

*Neurospora crassa* is the primary model species used to elucidate the circadian clock in filamentous fungi. Circadian rhythms in fungi are chiefly maintained by autoregulatory positive and negative feedback loops dictated by the FREQUENCY (FRQ)-FRQ-interacting RNA helicase (FRH) complex (FFC) [2, 19], which forms the negative arm, and the White Collar Complex (WCC): WHITE COLLAR-1 (WC-1, which also acts as a blue light photoreceptor) and WHITE COLLAR-2 (WC-2), forming the positive arm [16, 20]. Late in the circadian evening, the WCC binds to the *frq* promoter; *frq* levels then peak in the early morning and FRQ protein accumulates in the cytoplasm [2]. Here, some FRQ becomes partially phosphorylated by the clock component Ck1a, which then binds to FRH, creating the FRQ-FRH complex (FFC) [2, 19]. The FFC is then ubiquitinated by FWD-1 and is transported to the proteasome for turnover, which occurs throughout the day [2, 16]. Concurrently, some FRQ forms homodimers and moves into the nucleus [21]. The FRQ homodimer then represses *frq* transcription via its physical interaction with the WCC, preventing the WCC from acting as a *frq* transcription factor [2]. The FRQ levels fall as a result of turnover in the proteasome and its repression of transcription, and the WCC can again interact with the *frq* promoter region [2]. Nuclear FRQ also causes WC-1 to be synthesized, which activates *frq* transcription, contributing positive feedback and beginning the cycle again. This core clock is upstream of several other gene pathways, causing daily oscillations in expression and therefore controlling many physiological outputs in a time-dependent manner [1, 8, 11, 22].

The ubiquity of circadian rhythms suggests that it brings about a fitness advantage [2, 6, 15, 16, 23, 24]. In *N. crassa* and *M. oryzae*, conidiation occurs during the dark period, peaking just before dawn [16, 25]. At this time, temperatures are usually lower, and humidity tends to be higher than during the day, which may allow for optimal spore dispersal by wind and dewdrop early in the morning [25, 26]. Rhythmic conidiation corresponding to this natural behavior is seen in laboratory cultures of *M. oryzae* grown in a light-dark photoperiod (Fig. 1), which gives rise to a characteristic banding pattern on petri dishes.

Rhythmic conidiation is maintained in some fungal species when cultures are transferred to free-running conditions, and thus it may be considered a circadian output. Additionally, different wavelengths of light play more or less important roles in circadian outputs such as rhythmic conidiation [25] (Fig. 2); light wavelength composition and ratios differ depending on time of day, season, etc. – for example, at dawn and dusk, there is a much higher amount of red light compared to blue, giving us sunrise and sunset [11].

Conversely, circadian rhythms allow plants to anticipate timing of infection and prime their defenses accordingly [13, 27]. *Arabidopsis* exhibits circadian variation in susceptibility to both the bacterial pathogen *Pseudomonas syringae* and fungal necrotroph

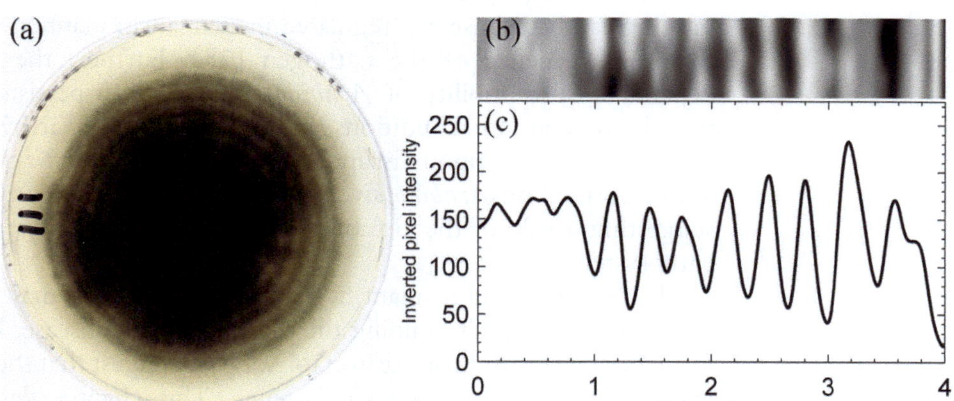


**Fig. 1** *Magnaporthe oryzae* rhythmic conidiation banding analysis. (**a**) *Magnaporthe oryzae* grown under 12 h/12 h light/dark cycling conditions for 14 days; dark bands of conidia are produced during the dark period. (**b**) Processed image transect of the banding pattern. (**c**) Output graph denoting colony radius, number of conidial bands, and distance between conidial bands

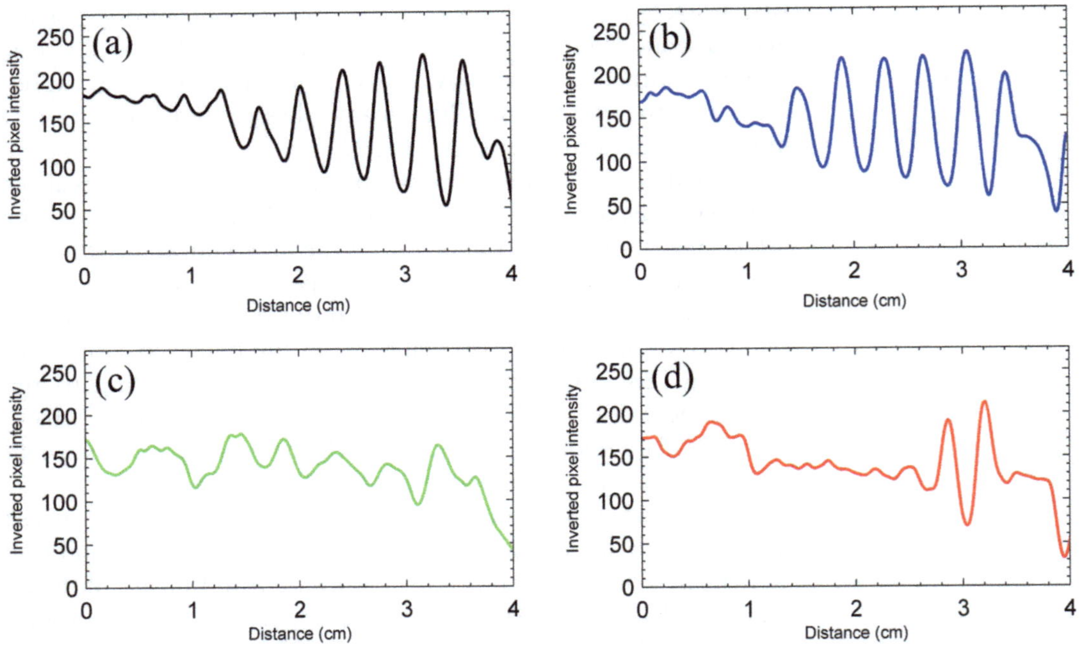


**Fig. 2** Wavelength-dependent conidial banding in *Magnaporthe oryzae*. Colonies grown under filters permitting different wavelengths of light: (**a**) white, (**b**) blue, (**c**) green, or (**d**) red light. Conidial banding is chiefly controlled by blue light

*Botrytis cinerea*, where *Arabidopsis* is most susceptible at dusk [28, 29]. There is a cyclic regulation of PAMP-triggered immunity, and when core clock genes are knocked out in *Arabidopsis* (producing an arrhythmic line), there was no difference in disease response to *Pseudomonas*, regardless of subjective inoculation time, suggesting the plant's defense was regulated in a circadian manner, rather than the virulence of the pathogen [28]. Further, the dusk-dependent susceptibility of *Arabidopsis* to *Botrytis* persists after transfer to constant conditions and is eradicated in arrhythmic mutants. However, upon disruption of the clock in *Botrytis cinerea*, lesions formed on *Arabidopsis* are significantly smaller than usual, suggesting a role of the circadian clock for both partners of the pathosystem [29].

In *M. oryzae* and its plant hosts, circadian rhythms may play a substantial role in determining the outcome of the disease. Bioinformatic analyses and comparison to *N. crassa* suggest that there are many homologous clock-related genes. This chapter describes methods used to analyze diurnal and circadian behavior shown in *Magnaporthe oryzae* to elucidate the function of the clock in the fungal pathogen.

## 2 Materials

### 2.1 Plant Material

1. Rice (CO-39 and Maratelli) and barley (Octavia and Golden Promise) seedlings are grown in growth chambers at 25 °C under 12 h of white light (~100μMol/$m^2$/s) and 12 h of dark for 2–3 weeks prior to infection.

### 2.2 Fungal Material

1. Fungal stocks (Guy11, kindly provided by the Nicholas Talbot lab, The Sainsbury Lab, Norwich) are grown from dried filter paper pieces penetrated by fungal mycelia. Filter stocks can be dried and stored at −20 °C in glassine bags inside a box with desiccant until needed.

### 2.3 Media

1. Complete medium (CM) [30]. To make 1 L, mix 50 mL of 20× nitrate salts (see below), 1 mL 1000× trace elements (see below), 10 g D-glucose, 2 g peptone, 1 g yeast extract, 1 g casamino acids, and 1 mL vitamin solution. Bring volume up to 1 L with double distilled water and adjust pH to 6.5 using NaOH. Add 15 g agar for solid media. Autoclave and then incubate at ~50 °C to keep molten. Add 50 mg penicillin and 50 mg streptomycin (from a 50 mg/mL stock solution; *see* **item 5**) per liter of media to prevent bacterial contamination. Pour media into 90 mm plates.
2. 20× nitrate salts. To make 1 L, use 120 g $NaNO_3$, 10.4 g KCl, 10.4 g $MgSO_4{\cdot}7H_2O$, and 30.4 g $KH_2PO_4$. Dilute to 1 L with double distilled water.
3. 1000× trace elements. To make 100 mL, use 2.2 g $ZnSO_4{\cdot}7H_2O$, 1.1 g $H_3BO_3$, 0.5 g $MnCl_2{\cdot}4H_2O$, 0.5 g $FeSO_4{\cdot}7H_2O$, 0.17 g $CoCl_2{\cdot}6H_2O$, 0.16 g $CuSO_4{\cdot}5H_2O$, 0.15 g $Na_2MoO_4{\cdot}2H_2O$, and 5 g $Na_4EDTA$. Add compounds in order in a beaker of double distilled water to dissolve and boil in the microwave. Cool to 60 °C and adjust pH to 6.5 (with KOH). Cool to room temperature and dilute to 100 mL with double distilled water.
4. Vitamin solution. To make 100 mL, use 0.01 g biotin, 0.01 g pyridoxine, 0.01 g thiamine, 0.01 g riboflavin, 0.01 g p-aminobenzoic acid (PABA), and 0.01 g nicotinic acid. Dilute to 100 mL with double distilled water. Store in a dark glass bottle at 4 °C.
5. Penicillin-streptomycin stock solution. Dissolve 500 mg of penicillin and 500 mg streptomycin sulfate into 10 mL sterile water. Filter sterilize and aliquot 500μL into sterile Eppendorf tubes. Store at −20 °C. Add 100μL stock solution per 100 mL of media.

## 3 Methods

### 3.1 Plant and Fungal Culture

#### 3.1.1 Plant Culture

1. Sterilize rice and barley seeds under 70% ethanol for 5 min. Replace the ethanol with water and wait a further 5 min—repeat this step twice.
2. Place seeds in a petri dish with damp paper in a 25 °C incubator under 12 h/12 h white light/dark.
3. After seeds have germinated (~3–4 days), place them into pots with compost and 20% ceramic granules.
4. Once transferred to pots, in the first week plants are watered once with nitrogen feed (Baby Bio) and once with micronutrients (Vitalink) according to the manufacturer's instructions. From the second week onward, plants are watered three times a week with tap water.
5. Rice and barley seedlings are then grown in growth chambers at 25 °C under 12 h of white light and 12 h of dark for 2–3 weeks total prior to inoculation.
6. Just before inoculation, live plants are placed in translucent plastic boxes, with damp paper towels attached to the side.
7. The lid of the box containing inoculated plants is removed after 2 days and plants are watered three times weekly. Water is added to the damp towels on the sides of the box.

#### 3.1.2 Fungal Culture

1. When seedlings are 1 week old, begin culturing *M. oryzae* colonies onto CM plates under experimental conditions.
2. Place a small piece of filter stock (stored at −20 °C) or a small agar plug from a previously growing colony (less than 12 days old) onto a CM plate and place into a 25 °C incubator under the chosen lighting regime.
3. Colonies can be grown under a number of different lighting regimes to assess diurnal/circadian effects on pathogenicity (*see* Subheading 3.2).
4. Fungal cultures are routinely grown either in-phase with plants, where dawn for plants is coincident with dawn for fungal cultures, or in antiphase, where dawn for plants is 12 h later than dawn for fungal cultures.

### 3.2 Diurnal and Circadian Culture and Analyses of Magnaporthe oryzae

#### 3.2.1 Diurnal Culture of Magnaporthe oryzae

1. Place a small piece of filter stock, a few mm in width, onto a CM petri dish.
2. Transfer colony to a 25 °C incubator under a chosen lighting regime; this is usually 12 h light/12 h dark. Colonies can be grown under different LED lights or under filters to assess the role of certain wavelengths of light on diurnal growth (Fig. 2).
3. After ~10 days, remove plates from incubators for analyses.
4. Time-lapse photography can be implemented to collect data throughout the colony growth.

#### 3.2.2 Circadian Culture of Magnaporthe oryzae

1. Place a small piece of filter stock, a few mm in width, onto a CM petri dish.
2. Transfer petri dish to a 25 °C incubator under a chosen lighting regime; commonly 12 h light/12 h dark. Colonies can also be grown under different LED lights or filters to assess the role of certain wavelengths of light on diurnal growth (Fig. 2).
3. Once the first conidial bands begin to appear (usually 6–8 days), transfer colonies to free-running conditions (continuous light or dark) for at least 2 days.
5. After ~10 days of total colony growth, remove plates from incubators for analyses.
6. Time-lapse photography can be implemented to collect data throughout the colony growth.

#### 3.2.3 Magnaporthe oryzae Banding Analysis

1. Place colonies upside down onto a light box to better differentiate conidial bands.
2. Photograph colonies.
3. Using an ImageJ macro (Supplementary Material 1), colony diameter (*M. oryzae* grows radially), number of conidial bands, and distance between conidial bands can be determined and plotted (Fig. 1). Using this information and because *M. oryzae* grows at a relatively consistent rate, periodicity can be calculated using:

$$\text{Period} = \frac{\text{Distance between conidial bands (mm)}}{\text{Average distance grown per hour (mm/h)}}.$$

   Alternatively, several software packages have been designed to interpret circadian data (*see* **Note 1**) [31, 32, 36–38].

4. Period, amplitude, and growth rate may all be affected by different incubation regimes, so relevant controls must be implemented.

### 3.3 Conidial Suspension Preparation, Conidial Analysis, and Plant Inoculations

#### 3.3.1 Conidial Suspension Generation and Plant Inoculations

1. Take an 8- to 12-day-old colony and add 3 mL double distilled, sterile water onto it.
2. Using a sterile spreader, spread water gently over the colony and knock off conidia until the colony no longer looks fluffy.
3. Pipette all remaining water containing conidia from the plate and filter the conidial suspension through one layer of Miracloth into a 1.5 mL Eppendorf tube.
4. Place the Eppendorf tube into a microfuge and spin at 3000 × *g* for 3 min.
5. Remove as much water as possible, leaving the dark conidial pellet behind.
6. Fill the Eppendorf with new sterile water to 1 mL and gently resuspend until the pellet has completely disappeared.
7. Determine the spore concentration under a hemocytometer and dilute to a concentration of $1 \times 10^5$ conidia/mL for spray inoculations.
8. For the final inoculation suspension, dissolve Tween 80 and gelatine to a final concentration of 0.1%.
9. Using an artist's airbrush, inoculate plants with the prepared spore suspension in large plastic boxes, ~3–5 mL suspension per pot of plants.
10. Secure the lid of the box to increase humidity and transfer to a plant growth chamber.
11. Continue watering plants as usual, and after 2 days remove the lid from the box.
12. Six days post-inoculation, leaves should be taped to a flatbed scanner and imaged for further analyses in disease severity.

#### 3.3.2 Conidial Counts to Determine Spore Tip Mucilage Production

This method is an altered version of Stanley et al. [31] and is used to determine spore tip mucilage production.

1. Harvest spores using the same technique as in Subheading 3.1, **step 3** and dilute to a concentration of $4 \times 10^5$ conidia/mL. Do not add gelatine or Tween 80.
2. Pipette 200μL of the spore suspension onto a 55 mm polystyrene petri dish.
3. Draw a line around the perimeter of the droplet using a permanent marker.
4. Incubate the petri dish for 1 h in an incubator at 25 °C.
5. Place under a microscope and capture at least three images of the spore suspension using a camera attachment.
6. Add 5 mL sterile water to the petri dish and seal lids with Parafilm.

7. Place in an orbital shaker for 100 s at 200 rpm.
8. Place under a microscope and capture a further three images of the spore suspension.
9. Using an ImageJ macro (Supplementary Material 2), conidial counts can be determined. Calculate the percentage spore remaining after agitation.

#### *3.3.3 Conidial Germination and Development Assay*

1. Harvest spores using the same technique as in Subheading 3.1, **step 3** and dilute to a concentration of $1 \times 10^5$ conidia/mL (manual assay) or $5 \times 10^5$ conidia/mL (time-lapse microscopy assay). Do not add gelatine or Tween 80.
2. Monitor spore germination and development either (a) manually or (b) using a time-lapse imaging microscope (Fig. 3).
   (a) *Manual development assay*:
   - Pipette 50μL of the spore suspension onto two cover slips per desired time point.
   - Place cover slips into a translucent box on top of damp paper towels and place under constant light in a 25 °C incubator.
   - Every 2 h for 24 h, remove two cover slips per time point and place on a slide under a microscope.
   - Count at least 100 spores per cover slip and note their developmental stage. In this method, completely new conidia are analyzed at every time point.

   (b) *Time-lapse microscopy development assay*:
   - Pipette a few mL of water into either an environmental chamber or 30 mm clear petri dish around the perimeter of the container to maintain humid conditions.
   - Pipette 50μL of the spore suspension onto either a slide in an environmental chamber or the center of a 30 mm petri dish.
   - Place under a time-lapse imaging microscope inside a 25 °C incubator and capture images every 30 min for 24 h (*see* **Notes 2** and **3**) [33].
3. Spore developmental stages are defined in [34, 35].

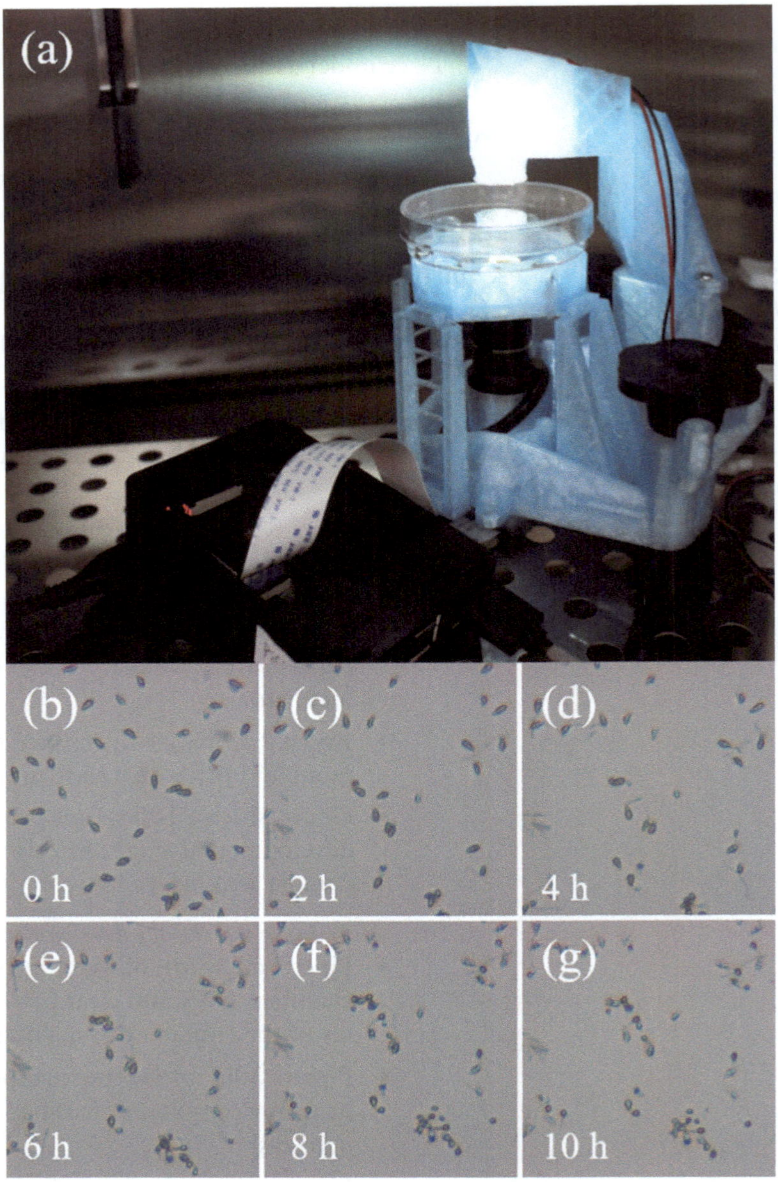


**Fig. 3** Time-lapse imaging of *Magnaporthe oryzae* conidial development. (**a**) 3D-printed OpenFlexure microscope connected to a Raspberry Pi computer that (**b–g**) captures time-lapse images of conidia every 30 min until prompted to stop

## 4 Notes

1. There are many software packages used for the analysis of circadian outputs to determine period and phase and generate actograms. Commonly used package sources are as follows:
   (a) Chronomics analysis toolkit (CATkit, actograms, period, phase) [32, 36].

(b) Circadian rhythm laboratory (actograms, period, phase response) [37].

(c) Excel (actograms) [38].

2. In our lab, we have used a 3D-printed microscope small enough to fit in an incubator (OpenFlexure [33]) connected to a Raspberry Pi computer that will automatically capture images every 30 min until prompted to stop.
3. A higher concentration of conidia is used in this method due to being constricted to one field of view, but the advantage is that the same spores are being imaged throughout the entirety of the experiment.

## References


1. Deng W, Clausen J, Boden S, Oliver SN, Casao MC, Ford B, Anderssen R, Trevaskis B (2015) Dawn and dusk set states of the circadian oscillator in sprouting barley (*Hordeum vulgare*) seedlings. PLoS One 10(6):e0129781
2. Baker CL, Loros JJ, Dunlap JC (2012) The circadian clock of *Neurospora crassa*. FEMS Microbiol Rev 36(1):95–110
3. Goodspeed D, Chehab EW, Min-Venditti A, Braam J, Covington MF (2012) Cozzarelli Prize Winner: *Arabidopsis* synchronizes jasmonate-mediated defense with insect circadian behavior. Proc Natl Acad Sci 109 (12):4674–4677
4. Dunlap J, Loros J, Colot H, Mehra A, Belden W, Shi M, Hong C, Larrondo L, Baker C, Chen C, Schwerdtfeger C, Collopy P, Gamsby J, Lambreghts R (2007) A circadian clock. In *Neurospora*: how genes and proteins cooperate to produce a sustained, entrainable, and compensated biological oscillator with a period of about a day. Cold Spring Harb Symp Quant Biol 72(1):57–68
5. Liu Y, Bell-Pedersen D (2006) Circadian rhythms in *Neurospora crassa* and other filamentous fungi. Eukaryot Cell 5 (8):1184–1193
6. Bodenstein C, Heiland I, Schuster S (2012) Temperature compensation and entrainment in circadian rhythms. Phys Biol 9(3):036011
7. McClung CR (2006) Plant circadian rhythms. Plant Cell 18(4):792–803
8. Bell-Pedersen D, Garceau N, Loros J (1996) Circadian rhythms in fungi. J Genet 75 (3):387–401
9. Más P, Yanovsky MJ (2009) Time for circadian rhythms: plants get synchronized. Curr Opin Plant Biol 12(5):574–579
10. Wang W, Barnaby JY, Tada Y, Li H, Tör M, Caldelari D, Dong X (2011) Timing of plant immune responses by a central circadian regulator. Nature 470(7332):110–114
11. Schumacher J (2017) How light affects the life of *Botrytis*. Fungal Genet Biol 106:26–41
12. Ingle R, Stoker C, Stone W, Adams N, Smith R, Grant M, Carré I, Roden L, Denby K (2015) Jasmonate signalling drives time-of-day differences in susceptibility of *Arabidopsis* to the fungal pathogen *Botrytis cinerea*. Plant J 84(5):937–948
13. Hua J (2013) Modulation of plant immunity by light, circadian rhythm, and temperature. Curr Opin Plant Biol 16(4):406–413
14. Roden L, Ingle R (2009) Lights, rhythms, infection: the role of light and the circadian clock in determining the outcome of plant-pathogen interactions. Plant Cell 21 (9):2546–2552
15. Pruneda-Paz JL, Kay SA (2010) An expanding universe of circadian networks in higher plants. Trends Plant Sci 15(5):259–265
16. Hevia MA, Canessa P, Larrondo LF (2016) Circadian clocks and the regulation of virulence in fungi: getting up to speed. Semin Cell Dev Biol 57:147–155
17. Baker CL, Dunlap JC (2010) Circadian rhythms: phosphorylating the CLOCK. Cell Cycle 9(2):227–232
18. Larrondo LF, Olivares-Yanez C, Baker CL, Loros JJ, Dunlap JC (2015) Decoupling circadian clock protein turnover from circadian period determination. Science 347 (6221):1257277–1257277
19. Cheng P, He Q, He Q, Wang L, Liu Y (2005) Regulation of the *Neurospora* circadian clock

by an RNA helicase. Genes Dev 19 (2):234–241

20. Hurley JM, Loros JJ, Dunlap JC (2016) The circadian system as an organizer of metabolism. Fungal Genet Biol 90:39–43
21. De Paula RMD, Lewis ZA, Greene AV, Seo KS, Morgan LW, Vitalini MW, Bell-Pedersen D (2006) Two circadian timing circuits in *Neurospora crassa* cells share components and regulate distinct rhythmic processes. J Biol Rhythm 21(3):159–168
22. Smolen P, Baxter DA, Byrne JH (2001) Modeling circadian oscillations with interlocking positive and negative feedback loops. J Neurosci 21(17):6644–6656
23. Faure S, Turner A, Gruszka D, Christodoulou V, Davis S, von Korff M, Laurie D (2012) Mutation at the circadian clock gene EARLY MATURITY 8 adapts domesticated barley (*Hordeum vulgare*) to short growing seasons. Proc Natl Acad Sci 109 (21):8328–8333
24. Filichkin SA, Breton G, Priest HD, Dharmawardhana P, Jaiswal P, Fox SE, Mockler TC (2011) Global profiling of rice and poplar transcriptomes highlights key conserved circadian-controlled pathways and cis-regulatory modules. PLoS One 6(6): e16907
25. Lee K, Singh P, Chung W, Ash J, Kim T, Hang L, Park S (2006) Light regulation of asexual development in the rice blast fungus, *Magnaporthe oryzae*. Fungal Genet Biol 43 (10):694–706
26. Deng Y, Qu Z, Naqvi N (2015) Twilight, a novel circadian-regulated gene, integrates phototropism with nutrient and redox homeostasis during fungal development. PLoS Pathog 11 (6):e1004972
27. Lu H, McClung CR, Zhang C (2017) Tick tock: circadian regulation of plant innate immunity. Annu Rev Phytopathol 55 (1):287–311
28. Bhardwaj V, Meier S, Petersen LN, Ingle RA, Roden LC (2011) Defence responses of *Arabidopsis thaliana* to infection by *Pseudomonas syringae* are regulated by the circadian clock. PLoS One 6(10):e26968
29. Hevia MA, Canessa P, Müller-Esparza H, Larrondo LF (2015) A circadian oscillator in the fungus *Botrytis cinerea* regulates virulence when infecting *Arabidopsis thaliana*. Proc Natl Acad Sci 112(28):8744–8749
30. Talbot NJ, Ebbole DJ, Hamer JE (1993) Identification and characterization of *MPG1*, a gene involved in pathogenicity from the rice blast fungus *Magnaporthe grisea*. Plant Cell 5 (11):1575
31. Zielinski T, Moore AM, Troup E, Halliday KJ, Millar AJ (2014) Strengths and limitations of period estimation methods for circadian data. PLoS One 9(5):e96462
32. Lee Gierke C, Cornelissen G (2015) Chronomics analysis toolkit (CATkit). Biol Rhythm Res 47(2):163–181
33. Collins J, Knapper J, Stirling J, Mduda J, Mkindi C, Mayagaya V, Mwakajinga G, Nyakyi P, Sanga V, Carbery D, White L, Dale S, Lim Z, Baumberg J, Cicuta P, McDermott S, Vodenicharski B, Bowman R (2020) Robotic microscopy for everyone: the OpenFlexure microscope. Biomed Opt Express 11:2447–2460
34. Hamer J, Howard R, Chumley F, Valent B (1988) A mechanism for surface attachment in spores of a plant pathogenic fungus. Science 239(4837):288–290
35. Inoue K, Suzuki T, Ikeda K, Jiang S, Hosogi N, Hyong G, Hida S, Yamada T, Park P (2007) Extracellular matrix of *Magnaporthe oryzae* may have a role in host adhesion during fungal penetration and is digested by matrix metalloproteinases. J Gen Plant Pathol 73(6):388–398
36. Chronomics Analysis Toolkit (2020) https://564394709114639785.weebly.com/. Accessed 10 May 2020
37. Circadian.org (2020) Circadian Rhythm Laboratory Software. https://www.circadian.org/softwar.html. Accessed 10 May 2020
38. Oike, Ogawa and Oishi (2019) Simple and quick visualization of periodical data using microsoft excel. Methods Protoc 2(4):81

# Chapter 14

# Methods to Study Autophagocytosis in *Magnaporthe oryzae*

**Lin Li, Xue-Ming Zhu, Xiao-Hong Liu, and Fu-Cheng Lin**

## Abstract

Autophagy is an evolutionarily conservative biological process in eukaryotes. Since the lysosomes were discovered by De Duve in the 1950s, autophagy has been studied for more than half a century and the mechanism of autophagy process has been discovered in many model organisms. In the rice blast fungus *Magnaporthe oryzae*, autophagy relative proteins are essential for appressorium formation, penetration, and invasive growth. The null mutants for the expression of autophagy gene homologs in *M. oryzae* lose their pathogenicity for infection of host plants. In this chapter, we provide some methods for monitoring autophagy process using physics and biochemistry assays in *M. oryzae*. Moreover, similar approaches can be used to monitor autophagy in other plant filamentous pathogenic fungi.

**Key words** Autophagy, *Magnaporthe oryzae*, Pathogenic fungi, GFP-Atg8, Atg8-PE, Autophagosome, TOR

## 1 Introduction

*Magnaporthe oryzae* is a model organism to study the interaction between plants and pathogens. It is a filamentous plant pathogenic fungus that can infect various crops such as rice and barley and cause rice blast and wheat blast [1, 2]. Appressoria produced by *M. oryzae* are specialized infectious structures that can penetrate the host plant cells. Appressorium formation is caused by fungal conidia attaching to the surface of the plant epidermis. A germ tube is produced from the attached conidium, and this germ tube grows, stimulated by the environment, thereby activating the intracellular molecular mechanism to form a mature appressorium [3]. Mature appressoria produce mechanical forces of up to 8 MPa, which allow penetration pegs to penetrate the plant surface [4].

Autophagy is a ubiquitous process in eukaryotes and plays an essential role in the normal life activities [5]. The main function of autophagy is to participate in the degradation of damaged proteins or organelles such as damaged mitochondria, peroxisomes, excess ribosomes, endoplasmic reticulum and endosomes, lipid droplets,

Stefan Jacob (ed.), *Magnaporthe oryzae: Methods and Protocols*, Methods in Molecular Biology, vol. 2356,
https://doi.org/10.1007/978-1-0716-1613-0_14, 

intracellular pathogens, and misfolded proteins [6]. Furthermore, in nutrient deficiency or environmental stress, autophagy can degrade excess organelles or proteins to provide new energy and metabolic substances for living [7]. Increasing evidences have been shown that autophagy pathways are essential for the normal growth and development of organisms. In mammals, loss of the autophagy core protein Atg5 causes death in embryonic mice [8]. About 40 proteins related to autophagy processes have been identified in *Saccharomyces cerevisiae*. These Atg proteins are divided into four groups: Atg1 kinase complex, ubiquitin-like Atg8/Atg12 coupling system, phosphatidylinositol 3-phosphate kinase complex, and Atg9 recycling complex. All of them participate in the entire autophagy process [9–12]. In *M. oryzae*, the autophagy pathway is closely related to the pathogenicity of the rice blast fungus [13]. Additionally, methods to study the autophagy process in *M. oryzae* are available, such as fluorescence localization of Atg proteins, biochemical methods, and detection of protein modifications through SDS-PAGE and Western blotting. In this chapter, we display details of all these methods.

## 2 Materials

### 2.1 GFP-MoAtg8 Fluorescence Localization

1. Complete agar medium (solid CM). To make 1 liter, mix 10 g glucose, 2 g peptone, 1 g yeast extract, 1 g casamino acids, 1 mL of trace elements, 1 mL of vitamin solution, 6 g $NaNO_3$, 0.5 g KCl, 0.5 g $MgSO_4$, 1.5 g $KH_2PO_4$, and 15 g agar, pH 6.5 [14].
2. Liquid complete medium (liquid CM). To make 1 liter, mix 10 g glucose, 2 g peptone, 1 g yeast extract, 1 g casamino acids, 1 mL of trace elements, 1 mL of vitamin solution, 6 g $NaNO_3$, 0.5 g KCl, 0.5 g $MgSO_4$, and 1.5 g $KH_2PO_4$, pH 6.5.
3. Vitamin solution. To make 100 mL, mix 0.01 g biotin, 0.01 g pyridoxine, 0.01 g thiamine, 0.01 g riboflavin, 0.01 g p-aminobenzoic acid (PABA), and 0.01 g nicotinic acid.
4. Trace elements. To make 100 mL, mix 2.2 g $ZnSO_4{\cdot}7H_2O$, 1.1 g $H_3BO_3$, 0.5 g $MnCl_2{\cdot}4H_2O$, 0.5 g $FeSO_4{\cdot}7H_2O$, 0.17 g $CoCl_2{\cdot}6H_2O$, 0.16 g $CuSO_4{\cdot}5H_2O$, 0.15 g $Na_2MoO_4{\cdot}5H_2O$, and 5 g $Na_4EDTA$.
5. SD-N liquid medium. To make 1 L, mix 20 g glucose and 1.7 g yeast synthetic dropout medium without amino acid 1.7 g [15].
6. *Magnaporthe* wildtype strain Guy11.
7. Nikon fluorescence microscopy.
8. CellTracker Blue (CMAC, 7-amino-4-chloromethylcoumarin).

### 2.2 Electron Microscopy for Monitoring Autophagosomes

1. Liquid complete medium (CM) (*see* **item 2** in Subheading 2.1).
2. Minimal medium without nitrogen (MM-N). To make 1 L, mix 10 g glucose, 0.5 g KCl, 0.5 g $MgSO_4$, 1.5 g $KH_2PO_4$, 1 mL of trace elements (see above), and 1 mL of vitamin solution (see above), pH 6.5. Add distilled water to 1 L and autoclave for 20 min.
3. Vitamin solution (*see* **item 3** in Subheading 2.1).
4. Trace elements (*see* **item 4** in Subheading 2.1).
5. The *M. oryzae* wildtype strain Guy11.
6. 2 mM phenylmethylsulfonyl fluoride (PMSF). Dissolve 0.348 mg PMSF in 1 mL methanol. Prepare in a fume hood. Store at −20°C.
7. 2% agarose. Weigh 2 g of agarose dissolved in 100 mL of 1× Tris–acetate-EDTA (TAE buffer) (see below).
8. 50×TAE buffer. Weigh 242 g of Tris and 18.612 g EDTA in a 1 L beaker. Add 800 mL of distilled water to the beaker, stir thoroughly, add 57.1 mL of glacial acetic acid, and dissolve fully. Adjust pH to 8.3 with NaOH, and finally add $ddH_2O$ to 1 L and store at room temperature. Dilute by 50 times when using a 1× TAE buffer.
9. Phosphate buffer (PBS). Weigh 71.628 g of $Na_2HPO_4{\cdot}12H_2O$ and add $ddH_2O$ to 1 L (solution B). Weigh 31.202 g of $NaH_2PO_4{\cdot}2H_2O$ and add $ddH_2O$ to 1 L (solution A). Add 305 mL of solution A and 195 mL of solution B to 500 mL of $ddH_2O$ to get 0.1 mol/L phosphate buffer solution. Use small amounts of solution A or B to adjust the pH to 7.0.
10. 2.5% glutaraldehyde solution. 50 mL of 25% glutaraldehyde stock solution, add 200 mL of $ddH_2O$, 150 mL of solution A, and 109 mL of solution B to dilute it to 2.5% glutaraldehyde solution. Prepare in a fume hood.
11. 1% citric acid. Add 1 mL of pure citric acid and dilute to 100 mL with PBS. Prepare in a fume hood (*see* **Note 12**).
12. 30% ethanol. Mix 30 mL of pure EtOH and 70 mL of $ddH_2O$; 30% refers to the volume fraction. 50%, 70%, 90%, and 95% EtOH also represent the volume fraction.
13. Spurr resin fixative. Pour 50 g of Nonenyl Succinic Anhydride (NSA) slowly into the beaker, add the last few drops with a spatula or stir bar, and add 20 g of ERL 4221 to the beaker with a disposable syringe. Then use a disposable syringe to add 16 g of DER 736 to the beaker. Finally, add 0.6 g of dimethylethanolamine (DMAE) to the beaker, remove from the balance, stir for about 3 min immediately, and place the beaker in a desiccator. The mixture releases bubbles and mixes well (*see* **Note 13**).

14. Lead citrate. Weigh 1.76 g of sodium citrate and 1.33 g of lead nitrate, add about 30 mL of dd$H_2O$ to the mark of the 50 mL volumetric flask, and shake for 30 min. The solution appears milky white. Add about 8 mL of 1 mol/L NaOH to clarify the solution. Prepare in a fume hood (*see* **Note 12**).
15. Uranyl acetate 50% ethanol saturated solution. Weigh 1.76 g of uranyl acetate, and add 25 mL of pure EtOH and 25 mL of dd$H_2O$. Then, add about 30 mL of dd$H_2O$ to the mark of the 50 mL volumetric flask.
16. Shake for 30 min. Prepare in a fume hood (*see* **Note 12**).

### 2.3 Western Blot to Detect the Degradation of GFP-MoAtg8 Protein and MoAtg8/MoAtg8-Phosphatidyleth anolamine) (PE) Turnover Assays

1. Liquid complete medium (CM) (*see* **item 2** in Subheading 2.1).
2. Vitamin solution (*see* **item 3** in Subheading 2.1).
3. Trace elements (*see* **item 4** in Subheading 2.1).
4. Lysis buffer. 50 mM Tris–HCl (pH 7.4), 150 mM NaCl, 1 mM EDTA, 1% Triton X-100. Autoclave for 20 min.
5. Protease inhibitor cocktail (Sigma-Aldrich, St. Louis, MO). Store at −20°C.
6. Phosphatase inhibitor cocktail 2 (Sigma-Aldrich). Store at 4°C.
7. *M. oryzae* wildtype strain Guy11.
8. 30% acrylamide (acrylamide-bis-acrylamide, 37.5:1) stock solution (Bio-Rad Laboratories, Hercules, CA).
9. 10% ammonium persulfate (APS) in dd$H_2O$.
10. Resolving gel buffer. 1.5 M Tris–HCl (pH 8.8), 0.4% SDS.
11. Resolving gel mixture. 4.2 mL of the 30% acrylamide stock, 2.5 mL of resolving gel buffer, 3.3 mL of dd$H_2O$, 45μL of 10% APS, and 10μL of *N,N,N,N′*-tetramethyl-ethylenediamine (TEMED).
12. Stacking gel buffer. 0.5 M Tris–HCl (pH 6.8), 0.4% SDS.
13. Stacking gel mixture. 0.75 mL of 30% acrylamide stock, 1.88 mL stacking buffer, 4.75 mL of dd$H_2O$, 40μL of 10% APS, and 15μL of TEMED.
14. Protein ladder (Thermo Fisher, No. 26616).
15. Bovine serum albumin (BSA).
16. SDS-PAGE running buffer. 25 mM Tris–HCl (pH 8.3), 200 mM glycine, 0.1% SDS.
17. 5× loading buffer. 100 mM Tris–HCl (pH 6.8), 4% SDS, 20% glycerol, 200 mM β-mercaptoethanol, 0.2% bromophenol blue.
18. Transfer buffer. 25 mM Tris–HCl (pH 8.5), 0.2 M glycine, 20% methanol.

19. 20× PBS (Sangon Biotech, Order No. B548117).
20. PBST. 1× PBS with 0.05% Tween 20.
21. Blocking solution. 1× PBST with 5% nonfat dry milk. Store at 4°C (*see* **Note 6**).
22. Antibody dilution buffer. 1× PBS with 5% BSA.
23. Anti-GFP antibody (Abcam; abE385).
24. Anti-ATG8 antibody (MBL; PM109).

### 2.4 Phosphorylation Assays to Monitor the Activity of TOR

1. Liquid complete medium (CM) (*see* **item 2** in Subheading 2.1).
2. Vitamin solution (*see* **item 3** in Subheading 2.1).
3. Trace elements (*see* **item 4** in Subheading 2.1).
4. *M. oryzae* wildtype strain Guy11.
5. 80% acetone.
6. 80% methyl alcohol with 0.5 M acetic acid.
7. 30% sucrose with 2% SDS.
8. 80% acetone with 10% trichloroacetic acid (TCA).
9. 8 M urea with 0.5% SDS.
10. 30% acrylamide (acrylamide-bis-acrylamide, 37.5:1) stock solution (Bio-Rad Laboratories, Hercules, CA).
11. 10% ammonium persulfate (APS) in $ddH_2O$.
12. Resolving gel mixture (5 mL). 1.33 mL of 30% acrylamide stock, 1.25 mL of 1.4 M Tris–HCl (pH 6.8), 50μL of 5 mmol phos binding reagent, 25μL of 20 mmol $ZnCl_2$, 2.35 mL of $ddH_2O$, 25μL of 10% APS, and 5μL of TEMED.
13. Stacking gel mixture (2 mL). 0.3 mL of 30% acrylamide stock, 0.5 mL of 1.4 M Tris–HCl (pH 6.8), 1.178 mL of $ddH_2O$, 10μL of 10% APS, and 2μL of TEMED.
14. Protein ladder (Thermo Fisher, No. 26616).
15. Bovine serum albumin (BSA).
16. SDS-PAGE running buffer. 25 mM Tris–HCl (pH 8.3), 200 mM glycine, 0.1% SDS.
17. 5× loading buffer. 100 mM Tris–HCl (pH 6.8), 4% SDS, 20% glycerol, 200 mM β-mercaptoethanol, 0.2% bromophenol blue.
18. Transfer buffer. 25 mM Tris–HCl (pH 8.5), 0.2 M glycine, 20% methanol.
19. 10× Tris-buffered saline (10× TBS). 0.5 M Tris–HCl (pH 7.4), 1.5 M NaCl.
20. TBST. 1× TBS with 0.05% Tween 20.

21. Blocking solution. 1× TBST with 3% nonfat dry milk. Store at 4°C.
22. Antibody dilution buffer. 1× TBST with 5% BSA.
23. Anti-FLAG antibody (Sigma).

## 3 Methods

### *3.1 GFP-MoAtg8 Fluorescence Localization*

1. In order to observe autophagy in hyphae, construct a vector containing GFP-MoAtg8 with Atg8 native promoter and transform into the wildtype Guy11 using *Agrobacterium tumefaciens*-mediated transformation (ATMT) [16].
2. Grow the strains on CM medium for 5 days.
3. Culture the strains in liquid CM medium for 2 days, then shift to SD-N medium for 4 or 8 h. Stain mycelia with CMAC for 30 min and obtain the photographs by fluorescence microscope (*see* **Note 1**).
4. Count the number of autophagosomes in each mycelium by ImageJ software (Fig. 1).
5. In order to monitor autophagy in conidia, grow the GFP-MoAtg8 strain on CM medium for 7 days.
6. Wash the conidia with dd$H_2O$ and dilute to $10^5$ conidia/mL.
7. Add 25μL on glass slide and culture at 25°C for 6, 12, 18, and 24 h.
8. Observe the number of autophagosomes in every conidium at per stage.
9. Analyze autophagosomes by image software.

### *3.2 Electron Microscopy for Monitoring Autophagosomes*

Carry out all procedures at room temperature unless noted with special temperature requirements.

1. Collect conidia of *M. oryzae* from 10-day-old cultures and then incubate them in the CM liquid medium at 28°C for 24 h in the incubator with shaking at 150 rpm (*see* **Note 2**).
2. Collect mycelium, wash it in distilled water thoroughly, and transfer it to MM-N liquid medium containing 2 mM PMSF. Incubate for 4 h at 28°C shaker at 150 rpm (*see* **Note 2**).
3. Embed mycelia that have been well incubated by 2% agarose.
4. Fix samples with 2.5% glutaraldehyde solution overnight at 4°C.
5. Discard the fixed solution and rinse the sample three times with a 0.1 mol/L PBS (pH 7.0) for 15 min each time.
6. Fix the sample with 1% $OsO_4$ for 1.5 h.

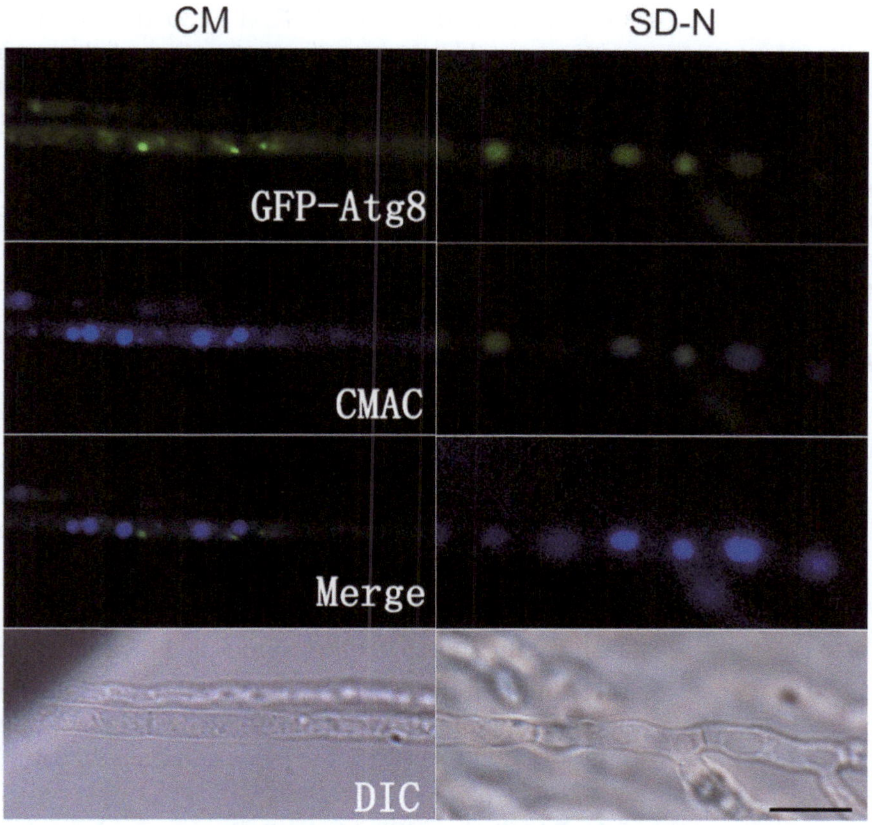


**Fig. 1** GFP-MoAtg8 localization in Guy11. Strains were grown in liquid CM medium at 25°C for 48 h and were shifted to liquid SD-N medium for 8 h. Mycelia were stained with CMAC and examined under a fluorescence microscope. Scale bar =10μm

7. As before, remove the fixative and rinse the sample three times with 0.1 mol/L PBS (pH 7.0) and each time 15 min.
8. Dehydrate the sample with ethanol dissolution with gradient concentrations including 30%, 50%, 70%, 80%, 90%, and 95% concentrations and treat the sample for 15 min at each concentration. Then treat with 100% ethanol for 20 min (*see* **Note 4**), Finally, treat it with pure acetone for 20 min.
9. Treat the samples with a mixture of Spurr resin fixative and acetone ($V/V = 1/1$) for 1 h.
10. Treat samples with a mixture of Spurr resin fixative and acetone ($V/V = 3/1$) for 3 h.
11. Embed the infiltrated sample and heat it overnight at 70°C to obtain an embedding sample.
12. Structural analysis. Section the samples on a Reichert ultramicrotome to obtain 70–90 nm sections.
13. Stain sections positively and stain with lead citrate and uranyl acetate 50% ethanol for 5 min (*see* **Notes 3** and **5**).

14. Observe it in the JEMOL JEM-1230 transmission electron microscope. Here, we perform a section of the strain Guy11 of *M. oryzae*, which is deficient in nitrogen and is treated with PMSF, and autophagy bodies are observed in vacuoles (Fig. 2).

#### 3.3 Western Blot to Detect the Degradation of GFP-MoAtg8 Protein and MoAtg8/MoAtg8-PE Turnover Assays

1. Protein extraction. Harvest hyphae from 5-day-old cultures of the wildtype strain Guy11. Culture the strain in liquid CM medium for 2 days, and shift to SD-N medium for 0, 6, and 12 h. Rinse the hyphae with sterile distilled water at room temperature. Remove excessive water by pressing hyphal pads between layers of paper towels. Grind mycelium with liquid nitrogen.
2. Transfer 300–500 mg of the semi-dry hyphae in a 2 mL screw cap microcentrifuge tube. Add 1 mL of protein lysis buffer (room temperature) containing 10μL of protease inhibitor cocktail and 10μL of phosphatase inhibitor cocktail 2 (*see* **Note 6**).
3. Centrifuge the lysate at 13,000 × *g* force for 20 min at 4°C.
4. Transfer the supernatant to a prechilled microcentrifuge tube.
5. Mix 200μL of the supernatant with 50μL of 5× loading buffer and denaturate the mixture in boiling water for 5 min. Centrifuge at 13,000 × *g* force for 1 min and keep the samples on ice (*see* **Note 7**).
6. SDS-PAGE for GFP-MoAtg8. Prepare polyacrylamide gels with resolving gel and stacking gel.
7. Load 10μL boiled protein sample in each well (*see* **Note 8**).

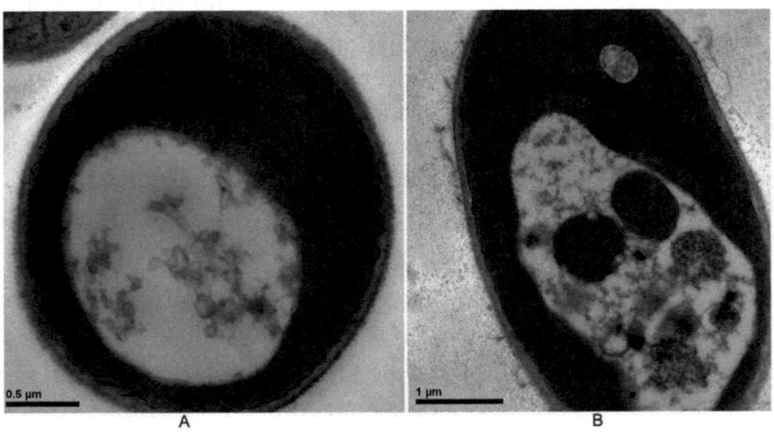


**Fig. 2** No autophagosomes were found in the vacuoles of wildtype Guy11 grown only on CM medium (**a**). After being shaken in CM medium and transferred to MM-N medium containing 2 mM PMSF and starvation culture consecutively for 4 h, autophagosomes appeared. Autophagosomes are indicated by arrows (**b**)

8. Run gel electrophoresis at 80 V until the protein markers are separated, then increase the voltage to 100–120 V.
9. Prepare polyvinylidene fluoride membranes and Whatman filter papers with the same dimensions of the SDS-PAGE gels. Lay one piece of nitrocellulose membrane over the SDS-PAGE gel and remove air bubbles by rolling with a glass rod. Lay two pieces of Whatman filter paper over the nitrocellulose membrane.
10. Place the gel polyvinylidene fluoride membrane package within the transfer cassette and transfer at 300 mA for 1–2 h (*see* **Note 9**).
11. Transfer the polyvinylidene fluoride membrane to 20 mL of blocking solution and incubate for 1 h at room temperature with gentle shaking.
12. Wash the membrane using 20 mL of TBST for 10 min with gentle shaking.
13. Repeat **steps 1–7** using the same loading, gel electrophoresis, and blotting conditions for Western blots: one for detecting GFP and GFP-Atg8 protein levels, one for detecting Atg8 and Atg8-PE levels.
14. For MoAtg8 and MoAtg8 turnover assays, we need to prepare 13.5% SDS-PAGE with 6 M urea.
15. Detection of protein levels. To determine the status of GFP and GFP-Atg8, add the anti-GFP antibody as the primary antibody at 1:10,000 dilution in antibody dilution buffer and incubate the samples overnight at 4°C with gentle shaking. To determine the status of Atg8-PE, use ATG8 antibody at a 1:2000 dilution.
16. Wash the membrane with 20 mL of TBST for 10 min at room temperature. Repeat three times.
17. Add the secondary antibody (anti-rabbit) at 1:2000 dilution in the antibody dilution buffer and incubate for 1 h at room temperature with gentle shaking.
18. Wash the membrane with 20 mL of PBST for 10 min at room temperature. Repeat three times.
19. Prepare a working solution of a hypersensitive chemiluminescent substrate by mixing equal volumes of solutions A and B. Mix the film thoroughly in the working solution and then expose it in the Bio-Rad ChemiDoc XRS+ machine.
20. Signals from the 27 and 41 kDa bands represent GFP and GFP-MoAtg8 (Fig. 3a), signal from 14 kDa represents MoAtg8, and the lower band next to MoATG8 is MoAtg8-PE (Fig. 3b).

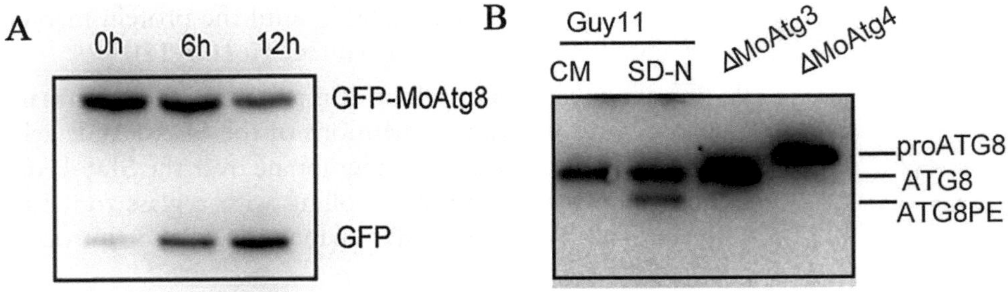


**Fig. 3** (**a**) Assays of GFP-PoAtg8 proteolysis in Guy11 at 25°C for 48 h in CM liquid medium with continuous shaking at 150 rpm. Autophagy was induced after 6 and 12 h of nitrogen starvation. Mycelia were collected at the indicated times, and mycelia extracts were analyzed by anti-GFP Western blotting. (**b**) The degradation of PoAtg8-PE is impaired in Guy11, *ΔMoatg3*, and *ΔMoatg4*. Strains were cultivated in CM liquid for 48 h and then shifted to nitrogen starvation medium (SD-N) for 8 h to induce autophagy. Immunoblot analysis of total lysates from the above hyphae, with anti-Atg8 antibody

### *3.4 Phosphorylation Assays to Monitor the Activity of TOR*

1. Construct the vector of Rps6-FLAG. Amplify about 1.2 kb RPS6 (RPS6 is 40S ribosomal protein S6-B, and its phosphorylation level can reflect the activity of TOR) sequence without terminal code by PCR as up flank, the flag sequence fuses with resistant gene as middle flank, and the 1.2 kb Rps6 terminal area sequence as down flank. Fuse three flanks with PKO1B [17] vector and transform them in *M. oryzae* by ATMT method. Verify the candidate strains by PCR as described previously.
2. Protein extraction (TCA-acetone-SDS method). Prepare 0.5–1 g of mycelium, grind with liquid nitrogen, and then add 2 mL of 80% acetone with 10% TCA. Incubate at −20°C for 2 h or overnight.
3. Centrifuge with 10,600 × *g* force for 10 min at 4°C; discard supplement.
4. Add 3 mL of 80% methyl alcohol with 0.5 M acetic acid, centrifuge with 10,600 × *g* force for 10 min at 4°C, and discard supplement (*see* **Note 10**).
5. Add 3 mL of 80% acetone, centrifuge with 10,600 × *g* force for 10 min at 4°C, discard supplement, and repeat this step once again (*see* **Note 10**).
6. Add 2 mL of 30% sucrose with 2% SDS, mix well and add 2 mL of Tris-saturated phenol, mix thoroughly and left at room temperature for 10 min, and centrifuge with 10,600 × *g* force for 10 min at 4°C.
7. Shift 1 mL of upper liquid and add 5 mL of 80% methyl alcohol with 0.5 M acetic acid. Incubate at −20°C for 4 h or overnight.
8. Centrifuge with 10,600 × *g* force for 10 min at 4°C, and discard supplement (*see* **Note 10**).

9. Wash the deposit twice with 3 mL of 80% acetone, discard supplement, and stand at room temperature for 10 min.
10. Add 0.5 mL of 8 M urea, mix well, and stay at room temperature for 10 min.
11. SDS-PAGE. Prepare polyacrylamide gels with resolving gel and stacking gel.
12. Load 10μL boiled protein sample in each well.
13. Run gel electrophoresis at 20 V for 5–6 h.
14. Wash the gel in a transfer buffer containing 10 mmol of EDTA for 10 min, repeat three times. Then wash the gel in a transfer buffer without EDTA for 10 min, repeat three times (*see* **Note 11**).
15. Prepare polyvinylidene fluoride membranes and Whatman filter papers with the same dimensions of the SDS-PAGE gels. Lay one piece of nitrocellulose membrane over the SDS-PAGE gel and remove air bubbles by rolling with a glass rod. Lay two pieces of Whatman filter paper over the nitrocellulose membrane.
16. Place the gel polyvinylidene fluoride membrane package within the transfer cassette and transfer at 300 mA for 3–5 h.
17. Transfer the polyvinylidene fluoride membrane to 20 mL of blocking solution and incubate for 30 min at room temperature with gentle shaking.
18. Wash the membrane using 20 mL of TBST for 10 min with gentle shaking.
19. Detection of protein levels. To determine the status of FLAG, add the anti-FLAG antibody as the primary antibody at 1:2000 dilution in antibody dilution buffer and incubate overnight at 4°C with gentle shaking.
20. Wash the membrane with 20 mL of TBST for 10 min at room temperature. Repeat three times.
21. Add the secondary antibody (anti-mouse) at 1:2000 dilution in the antibody dilution buffer and incubate for 1 h at room temperature with gentle shaking.
22. Wash the membrane with 20 mL of PBST for 10 min at room temperature. Repeat three times.
23. Prepare a working solution of a hypersensitive chemiluminescent substrate by mixing equal volumes of solutions A and B. Mix the film thoroughly in the working solution and then expose to X-ray film.
24. Signals from Phos-tag have two bands: p-Rps6 represents phosphorylated, and Rps6 represents non-phosphorylated; use normal SDS-PAGE bands as controls (Fig. 4).

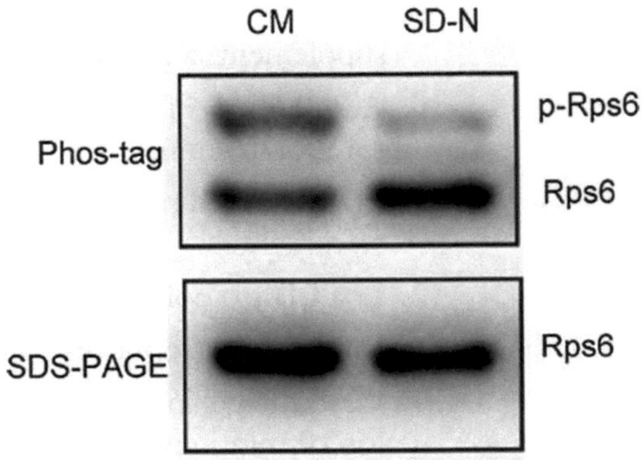


**Fig. 4** MoRps6-FLAG proteins treated with phosphatase inhibitor or phosphatase were separated by $Zn^{2+}$ Phos-tag SDS-PAGE and normal SDS-PAGE, respectively, and were probed with FLAG antibody

## 4 Notes

1. Choose the similar hyphae to observe and achieve statistics and use enough single hyphae.
2. When *M. oryzae* is transferred to CM or MM-N medium, it must be operated in a clean bench to prevent contaminations.
3. Store citric acid, lead citrate, and uranyl acetate in a fume hood when operating and protect from light.
4. It should be noted that when using alcohol with a concentration higher than 70%, timely cover the lid to prevent the alcohol from evaporating.
5. When dyeing with lead citrate, add a small amount of NaOH around the dye plate in order to prevent reactions from generating black particles.
6. 5% (w/v) bovine serum albumin (BSA) can be used as an alternative blocking reagent instead of milk. Add phosphatase inhibitors to lysis buffer immediately before use. To reduce or avoid proteolysis and dephosphorylation, keep the samples on ice as soon as lysis buffer is added.
7. Centrifuge the samples to remove insoluble tissue debris before running the SDS-PAGE.
8. It is important to load an identical amount of proteins in each lane for comparison.
9. Wet polyvinylidene fluoride in methanol for a few seconds before making the sandwich package will facilitate the protein transfer.
10. Pipette the supernatant carefully; do not mix into the supplement.

11. The concentration of EDTA in the transfer solution can be adjusted to 5–10 mmol. After washing with a transfer solution containing 10 mmol EDTA, be sure to wash three times with a normal transfer solution without EDTA.
12. Keep away from light.
13. Due to the long service life (3–4 days), the mixture can be prepared 1 day before use.

## References


1. Jeon J, Park SY, Chi MH, Choi J, Park J, Rho HS, Kim S, Goh J, Yoo S, Choi J, Park JY, Yi M, Yang S, Kwon MJ, Han SS, Kim BR, Khang CH, Park B, Lim SE, Jung K, Kong S, Karunakaran M, Oh HS, Kim H, Kim S, Park J, Kang S, Choi WB, Kang S, Lee YH (2007) Genome-wide functional analysis of pathogenicity genes in the rice blast fungus. Nat Genet 39(4):561–565
2. Talbot NJ (2003) On the trail of a cereal killer: exploring the biology of *Magnaporthe grisea*. Annu Rev Microbiol 57:177–202
3. Liu XH, Gao HM, Xu F, Lu JP, Devenish RJ, Lin FC (2012) Autophagy vitalizes the pathogenicity of pathogenic fungi. Autophagy 8 (10):1415–1425
4. Dagdas YF, Yoshino K, Dagdas G, Ryder LS, Bielska E, Steinberg G, Talbot NJ (2012) Septin-mediated plant cell invasion by the rice blast fungus, *Magnaporthe oryzae*. Science 336 (6088):1590–1595
5. Tooze SA, Yoshimori T (2010) The origin of the autophagosomal membrane. Nat Cell Biol 12(9):831–835
6. Hurley JH, Young LN (2017) Mechanisms of autophagy initiation. Annu Rev Biochem 86:225–244
7. Zhu XM, Li L, Wu M, Liang S, Shi HB, Liu XH, Lin FC (2019) Current opinions on autophagy in pathogenicity of fungi. Virulence 10(1):481–489
8. Arakawa S, Honda S, Yamaguchi H, Shimizu S (2017) Molecular mechanisms and physiological roles of Atg5/Atg7-independent alternative autophagy. Proc Jpn Acad Ser B Phys Biol Sci 93(6):378–385
9. Kim J, Huang WP, Klionsky DJ (2001) Membrane recruitment of Aut7p in the autophagy and cytoplasm to vacuole targeting pathways requires Autlp, Aut2p, and the autophagy conjugation complex. J Cell Biol 152 (1):51–64
10. Nair U, Cao Y, Xie Z, Klionsky DJ (2010) Roles of the lipid-binding motifs of Atg18 and Atg21 in the cytoplasm to vacuole targeting pathway and autophagy. J Biol Chem 285 (15):11476–11488
11. Nakatogawa H, Ishii J, Asai E, Ohsumi Y (2012) Atg4 recycles inappropriately lipidated Atg8 to promote autophagosome biogenesis. Autophagy 8(2):177–186
12. Noda NN, Fujioka Y (2015) Atgl family kinases in autophagy initiation. Cell Mol Life Sci 72(16):3083–3096
13. Talbot NJ, Kershaw MJ (2009) The emerging role of autophagy in plant pathogen attack and host defence. Curr Opin Plant Biol 12 (4):444–450
14. Talbot NJ, Ebbole DJ, Hamer JE (1993) Identification and characterization of *MPG1*, a gene involved in pathogenicity from the rice blast fungus *Magnaporthe grisea*. Plant Cell 5 (11):1575–1590
15. Suzuki SW, Onodera J, Ohsumi Y (2011) Starvation induced cell death in autophagy-defective yeast mutants is caused by mitochondria dysfunction. PLoS One 6(2):e17412
16. Liu XH, Chen SM, Gao HM, Ning GA, Shi HB, Wang Y, Dong B, Qi YY, Zhang DM, Lu GD, Wang ZH, Zhou J, Lin FC (2015) The small GTPase MoYpt7 is required for membrane fusion in autophagy and pathogenicity of *Magnaporthe oryzae*. Environ Microbiol 17 (11):4495–4510
17. Cao H, Huang P, Zhang L, Shi Y, Sun D, Yan Y, Liu X, Dong B, Chen G, Snyder JH, Lin F, Lu J (2016) Characterization of 47 Cys2 -His2 zinc finger proteins required for the development and pathogenicity of the rice blast fungus *Magnaporthe oryzae*. New Phytol 211(3):1035–1051

# Chapter 15

# Molecular Detection of *Magnaporthe oryzae* from Rice Seeds

Sonia Chadha


## Abstract

Rice blast disease caused by the fungus *Magnaporthe oryzae* is one of the most devastating diseases of rice worldwide. Blast pathogen infects all stages of rice causing leaf, collar, neck, and panicle blast symptoms. Seedlings infested by *M. oryzae* serve as an inoculum source, which gradually causes the disease symptoms on rice leaves. Hence, for the blast disease management, it is crucial to detect the pathogen in rice seeds, particularly at the presymptomatic stage. Early pathogen diagnosis enhances the accuracy and timing of fungicide applications, thereby improving their efficiency. Moreover, detection of infested seeds is important for the quarantine purposes to ensure the flow of quality rice seeds to the market. In this chapter, a PCR-based assay is described to detect the blast pathogen from rice seeds. The ability of this molecular method in reliably detecting pathogens can prevent the spread of blast disease because of its increased sensitivity and the reduction of diagnostic time.


**Key words** PCR detection, Pathogen, Molecular detection, Seed assay, Rice blast, *Magnaporthe oryzae*

## 1 Introduction

*Magnaporthe oryzae* causes rice blast, the most serious fungal disease of cultivated rice. The pathogen infects all aerial parts of rice and can survive in rice seeds, straws, and residues. The primary blast infection often occurs in empty glumes, pedicels, and distant ends of the hulls. Blast pathogen prefers the embryonal site of the rice seed for sporulation, as it allows the fungus to quickly access the shoot coming out from the seed [1]. The planting of infested rice seeds can introduce the fungus into new areas. Reports of the seed transmission of the blast pathogen from different parts of the world [2] suggest the significant economic impact of seed-borne blast infections, particularly in underdeveloped countries where chemical treatments of seed are limited and expensive [3]. Seeds appearing healthy may contain low inoculum of pathogen populations that

Stefan Jacob (ed.), *Magnaporthe oryzae: Methods and Protocols*, Methods in Molecular Biology, vol. 2356, https://doi.org/10.1007/978-1-0716-1613-0_15, © Springer Science+Business Media, LLC, part of Springer Nature 2021

later serves as an inoculum source, thereby resulting in the progressive disease development in the crop. Therefore, early *M. oryzae* detection in infested rice seeds is required for the pre- and postharvest management of blast disease. The conventional approaches for pathogen detection in seeds include mainly time-consuming agar plate and blotter methods [4]. Few studies also used serological assays for *M. oryzae* detection [5]. These methods are prone to cross-contamination, are time-consuming, and need experienced and skilled laboratory staff. These drawbacks of conventional methods can be overcome by the polymerase chain reaction (PCR) diagnostic assays, which are specific, sensitive, and rapid. In the PCR technique, nucleic acid is amplified in vitro by repeated cycles of denaturation of double-stranded DNA at high temperature and annealing of primers to the single-stranded template followed by DNA polymerase-mediated synthesis [6, 7]. The development and success of the PCR-based seed diagnostic assay mainly relies on three fundamental steps that include DNA target selection, designing of *M. oryzae*-specific primers, and quality of the DNA template. Commonly used DNA targets for diagnostic assays include single copy genes (beta-tubulin gene, calmodulin gene, or mating-type genes) and multiple copy genes (rDNA, mitochondrial, or repetitive DNA). However, single copy pathogenicity and infection-related genes are considered as ideal targets as they uniquely define the target pathogen and are thus preferred as target DNA for obtaining blast pathogen-specific primers [8]. Further, the efficiency of the PCR diagnostic assay relies on the primer selection, whereby an ideal primer pair should amplify DNA fragments from all diverse *M. oryzae* isolates. Several factors affect the specificity of primers such as annealing temperature, magnesium concentration, primer length, and secondary structure of target and primer sequences. Ideal oligonucleotides have 50% G + C content and are typically 18–22 nucleotides in length and without complementary 3′ ends or inverted repeats. Diagnostic primers that generate amplicons in the size range of 100–1000 bp are preferred as they can be efficiently PCR amplified and resolved by agarose gel electrophoresis. Furthermore, the DNA template quality also affects the efficiency of PCR assays.

Fungal pathogens are usually present at very low inoculum levels in seeds, especially in presymptomatic seeds. However, the low pathogen infestation in the seed cannot be ignored as it is capable of causing serious outbreaks of disease. The detection of the blast pathogen in rice seeds is extremely difficult due to these low inoculum levels [9]. This problem is resolved in the present method by including an enrichment step to increase the *M. oryzae* fungal biomass in the infested rice seeds. This step not only increases the sensitivity of a seed diagnostic assay but also reduces the possibility of generating amplicons from nonviable cells.

Moreover, PCR inhibitory substances such as seed phenolic compounds, polysaccharides, and proteins can be avoided by removing the mycelia from rice seeds prior to the DNA extraction step [8, 10].

The PCR diagnostic assay should have high specificity for the pathogen detection. *M. oryzae*-specific primers should generate a target signal in samples of diverse blast isolates, while the target signal should be absent in samples of nonspecific organisms such as other seed pathogens, saprophytes, or those related to the blast pathogen. Detection assays should also have high sensitivity, which depends primarily on the type of tissue used, the DNA extraction method, the presence of extraneous DNA, and PCR inhibitory substances. These challenges have been addressed in the present molecular seed diagnostic method for *M. oryzae*. The described PCR detection assay is simple and robust, with minimal risk of cross-contamination.

## 2 Materials

### 2.1 Equipment

1. Laminar flow bench.
2. Refrigerated orbital shaker.
3. Refrigerated centrifuge.
4. UV series spectrophotometer.
5. Thermal gradient cycler (Eppendorf Mastercycler® gradient thermal cycler or equivalent).
6. Horizontal gel electrophoresis apparatus.
7. Power pack.
8. Shaker for gel staining.
9. UV transilluminator with attached camera containing appropriate filter for visualization and photographing DNA gels.
10. Micropipettes: Variable volume pipettes, preferably dedicated for PCR and autoclavable.
11. Autoclaved microtubes (1.5, 2.0 mL).
12. Autoclaved thin-walled clear PCR tubes (200μL).
13. Autoclaved pipette microtips of 10, 200, and 1000μL. Pre-sterilized filter tips can also be used.
14. Miscellaneous items: Blotting paper, mortar and pestle, liquid nitrogen, spatula, muslin cloth, and glass funnel.

### 2.2 Media and Reagents

1. Potato dextrose broth (PDB) [(20% potato infusion and 2% dextrose) pH 5.6]. Prepare media in distilled water and autoclave at 121 °C for 15 min.

2. Potato dextrose agar (PDA): Add 2% agar to PDB medium and autoclave at 121 °C for 15 min.
3. DNA extraction buffer containing 100 mM Tris–Cl (pH 8.5), 250 mM NaCl, and 25 mM EDTA. Adjust pH to 8.0, autoclave, and store for up to 1 month at room temperature.
4. 3 M sodium acetate (pH 5.2). Weigh 24.6 g of sodium acetate (molecular weight = 82.03) and dissolve in double distilled water ($ddH_2O$). Adjust pH to 5.2 with glacial acetic acid and adjust final volume to 100 mL. Autoclave and store at 4 °C.
5. 5 M potassium acetate solution.
6. Tris–EDTA (TE) buffer (10 mM Tris–HCl, 1 mM EDTA, pH 8.0). Mix 1 mL of 1 M Tris–HCl buffer (pH 8.0), 200μL of 0.5 M EDTA (pH 8.0), and 98.8 ml of $ddH_2O$. Autoclave and store at 4 °C.
7. RNase A. Dissolve 20 mg of RNase A in 2 mL of 10 mM Tris–HCl (pH 8.0). Boil RNase A solution for 10 min and store as 200μL aliquots at −20 °C.
8. dNTP stock solution. 10 mM stock of dNTPs (containing dATP, dCTP, dGTP, and dTTP) in sterile $ddH_2O$. Prepare 100μL aliquots and store at −20 °C.
9. Oligonucleotide primers. 100μM stock solutions of primers in sterile DNase and RNase-free water. Prepare 10μM working stock and store as 100μL aliquots at −20 °C.
10. DNA gel loading dye (6×). Mix 25 mg of bromophenol blue, 25 mg of xylene cyanol, 3.3 mL of glycerol, and 6.7 mL of sterile 10 mM Tris–HCl (pH 8.0). Aliquot and freeze at −20 °C for long-term storage.
11. 10× Tris–borate-EDTA (TBE) buffer. Dissolve 121.1 g Tris–HCl, 61.8 g boric acid, and 7.4 g EDTA (disodium salt) in $H_2O$, adjust pH to 8.0, and make final volume to 1 L. Autoclave and store for up to 6 months at room temperature.
12. 10 mg/mL ethidium bromide stock. Dissolve 100 mg ethidium bromide in 10 mL of sterile distilled water.
13. 0.1% mercuric chloride in distilled water.
14. 10% SDS in distilled water.
15. 24:1 chloroform/isoamylalcohol solution.
16. 1:1 phenol/chloroform solution.
17. EtOH.
18. 10× *Taq* DNA polymerase buffer. 100 mM Tris–Cl (pH 9.0), 500 mM KCl, 15 mM $MgCl_2$, 0.1% gelatin.
19. Control primer pair ITS4 (5′- GGAAGGAGAAGTCGTAA CAAGG-3′) and (5′-TCCTCCGCTTATTGATATGC-3′).

20. *M. oryzae*-specific primer pair (MIF). MIF-forward (5′-GGAT CCATTGAGCATGCGTT-3′) and MIF-reverse (5′-GGATC CAATACGATCACTCG-3′). These primers are derived from *Magnaporthe* infection-specific 23 (*mif* 23) gene (GenBank accession no. GI 4732021).
21. Molecular weight DNA marker of 100–10,000 base range GeneRuler™ DNA Ladder Mix (Thermo Scientific) or similar. Dilute to 20 ng/μL for agarose gel electrophoresis.

## 3 Methods

### 3.1 Fungal Growth Conditions

1. Inoculate *M. oryzae* isolates and nonspecific fungi on PDA plates.
2. Grow at 25 °C for 5–8 days. For routine use, store culture plates at 4 °C.

### 3.2 Genomic DNA Isolation

1. Inoculate a hyphal plug from the PDA culture plate in 100 mL of PDB. Grow for 4–5 days at 25 °C at 100 rpm.
2. Filter mycelial mat through double layer of sterile muslin cloth. Rinse mycelial mat with 200 mL of sterile distilled $H_2O$ and immediately dry between sterile blotting papers. Grind dried mycelial mat in liquid nitrogen to fine dry powder using mortar and pestle (*see* **Note 1**).
3. Transfer around 100 mg powdered mycelia to a 2.0 mL microcentrifuge tube containing 500μL of pre-warmed DNA extraction buffer and 25μL of 10% SDS. Mix thoroughly and incubate at 65 °C for 30 min (*see* **Note 2**).
4. Add 170μL of 5 M potassium acetate solution to sample tubes. Mix and incubate on ice for 30 min.
5. Add 500μL of chloroform/isoamylalcohol (24:1). Gently mix and centrifuge for 10 min at 11,500 × *g* (*see* **Note 3**).
6. Remove aqueous (upper) layer and add equal volume of ice-cold isopropanol. Mix by inverting the tube. Incubate at room temperature for 20 min. Centrifuge at 11,500 × *g* for 10 min. Discard supernatant and dry the pellet in laminar flow (*see* **Note 4**).
7. Dissolve DNA pellet in 500μL of TE buffer (pH 8.0).
8. Add 5μL of 10 mg/mL RNase to the DNA sample and incubate at 37 °C for 1 h.
9. Add 500μL of phenol/chloroform solution to the sample. Mix and centrifuge for 5 min at 11,500 × *g* (*see* **Note 3**).
10. Transfer aqueous layer in a fresh tube and add 500μL of chloroform:/isoamylalcohol solution. Mix and centrifuge for 5 min at 11,500 × *g*. Repeat this step (*see* **Note 3**).

11. To aqueous layer, add 1/tenth volume of 3 M sodium acetate and two volumes of chilled EtOH. Mix and incubate at −20 °C for 30 min. Centrifuge at 11,500 × *g* for 10 min at 4 °C.
12. To pellet, add 100μL of 75% EtOH. Centrifuge at 11,500 × *g* for 5 min. Repeat this step. Air dry pellet at room temperature in laminar flow and dissolve in 100μL of 10 mM Tris–HCl buffer (pH 8.0) (*see* **Note 4**).
13. Quantify genomic DNA samples by recording DNA absorbance at 260 and 280 nm using the NanoQuant plate in the Infinite M200PRO Multimode Reader (Tecan, Austria) (*see* **Note 5**).

### 3.3 Development of M. oryzae-Specific Primers

1. To design *M. oryzae*-specific primers, perform literature search and choose pathogenicity-related gene unique to the blast pathogen. Obtain its DNA sequence from NCBI-nucleotide database (https://www.ncbi.nlm.nih.gov/nucleotide) or Uni-Prot (https://www.uniprot.org) database. In this protocol, *Magnaporthe* infection-specific 23 (*mif23*) gene is used to design MIF-forward and MIF-reverse primers (*see* **Note 6**).
2. Use the online tools such as Primer3 (http://bioinfo.ut.ee/primer3/ [11]) to design optimal PCR primers.
3. Select various parameters such as G + C content, primer melting temperature ($T_m$), product size range, and intra- and/or inter-complementarity of primers.
4. For effective PCR, calculate $T_m$ and properties of the designed primers using OligoCalc software (http://biotools.nubic.northwestern.edu/OligoCalc.html [12]).
5. Based on the aforementioned criteria, select primers for custom synthesis. Many different suppliers provide the service of oligonucleotide synthesis (*see* **Note 6**).

### 3.4 Control PCR and Agarose Gel Electrophoresis

1. Perform control PCRs using conserved fungal ITS primers (*see* **Note 7**).
2. Perform PCR reactions in 0.2 mL clear and thin-walled PCR tubes. Perform all PCR setup experiments on ice in a clean laminar flow bench (*see* **Note 8**).
3. PCR reaction: Mix 100μM dNTPs, 0.4μM of each primer ITS4 and ITS5 [13], *M. oryzae* genomic DNA (2 ng), 0.5 units of *Taq* DNA polymerase, and 2.5μL of 10× *Taq* DNA polymerase buffer. Add sterile water to make final volume of 25μL (*see* **Note 9**).
4. Perform PCR assay in a thermocycler using the following program: 94 °C for 5 min, 35 cycles of 92 °C for 30 s, 60 °C for 1 min, 72 °C for 1 min, and a final extension of 72 °C for 8 min.

5. Mix 20μL of PCR amplified products with 4μL of 6× gel loading dye and resolve on 1.5% agarose gel in 0.5× TBE buffer (pH 8.0) at 100 V using horizontal gel electrophoresis apparatus.
6. Use 100 bp DNA ladder as a molecular weight marker.
7. Stain gels with 0.5μg/mL ethidium bromide (*see* **Note 10**) for 45 min followed by destaining in distilled water for 10 min. Visualize DNA bands in agarose gels using UV transilluminator.

### *3.5 M. oryzae-Specific PCR Assay*

1. To obtain single and distinct pathogen-specific signals, first optimize PCR components by testing the range of concentrations of fungal genomic DNA (1–20 ng), dNTPs (25–200μM), specific primer pair MIF-forward and MIF-reverse (0.05–0.4μM), and *Taq* DNA polymerase (0.3–1.0 units) (*see* **Note 11**).
2. Optimize PCR conditions by varying denaturation temperature (92–5 °C), denaturation time (30–60 s), annealing temperature (55–65 °C), annealing time (30–60 s), and number of PCR cycles (*see* **Notes 11** and **12**).
3. The optimal amplification reaction mixture for the *M. oryzae*-specific primers used in this protocol: 50μM dNTPs in $ddH_2O$, 0.1μM of each primer, *M. oryzae* genomic DNA (5 ng), 0.5 units of *Taq* DNA polymerase, and 2.5μL of 10× *Taq* DNA polymerase buffer. Add sterile $ddH_2O$ to final volume of 25μL. Include negative control (without DNA template) in all the PCR experiments.
4. Perform PCR in a thermocycler using the following program: 94 °C for 5 min, 35 cycles of 92 °C for 30 s, 60 °C for 1 min, 72 °C for 1 min, and a final extension of 72 °C for 8 min.
5. Resolve PCR-amplified DNA using agarose gel electrophoresis as described in **steps 5–7** of Subheading 3.4.

### *3.6 Detection of M. oryzae in Infected Seeds*

1. Evaluate naturally *Magnaporthe*-infested rice seed lots for the presence and percentage of infestation of the rice blast pathogen. Surface sterilize the seeds by rinsing once with 0.1% mercuric chloride for 20 s followed by three times wash with sterile distilled water (1 min each).
2. Following surface sterilization, perform enrichment step by adding seeds to 10 mL of PDB media. Incubate for 48 h at 25 °C at 100 rpm (*see* **Note 13**).
3. Filter the contents through double layer of sterile muslin cloth and discard culture filtrate.
4. Collect mycelia growing on seeds for DNA extraction and quantification using the method described in Subheading 3.2.

5. Use purified genomic DNA (2 ng) with *M. oryzae*-specific primers for PCR. Perform PCR as described in Subheading 3.5.
6. Resolve PCR-amplified products as described in Subheading 3.4.

### 3.7 Specificity and Sensitivity Assays

1. Specificity assay: Perform PCR reaction using *M. oryzae*-specific primers to amplify target DNA band using genomic DNA (2 ng) from diverse *M. oryzae* isolates and nonspecific fungi. Confirm the presence of target DNA band in *M. oryzae* isolates and absence in the nonspecific fungi. Use amplification and electrophoresis conditions as described in Subheading 3.4 (*see* **Note 11**).
2. Sensitivity assay: To find the detection limits of *M. oryzae*-specific primers, perform PCR with known amounts of *M. oryzae* DNA (from 2 fg to 20 ng) and analyze. Perform PCR as described in **steps 3** and **4** of Subheading 3.5. Resolve PCR products using agarose gel electrophoresis as described in **steps 5–7** of Subheading 3.4 (*see* **Note 14**).
3. Sensitivity assay for infested seeds: To prepare artificially infested seed lots, mix the naturally infested seeds with noninfested rice seeds in the following proportions: 1:1 (50%), 1:20 (5%), 1:100 (1%), 1:500 (0.2%), and 1:1000 (0.1%). Perform seed evaluation assay as described in Subheading 3.6 (*see* **Notes 13** and **14**).

## 4 Notes

1. Cryogenic grinding using liquid nitrogen is a very effective technique for fungal mycelia. While performing cryogenic grinding, the fungal sample may warm up which can be avoided by adding liquid nitrogen. After grinding, proceed immediately to the next step to avoid thawing of fungal samples.
2. For lysis of fungal cells, use pre-warmed DNA extraction buffer (65 °C). Store SDS solution at room temperature as it precipitates at 4 °C.
3. Check the compatibility of labwares with chloroform and phenol. Store chloroform/isoamylalcohol and phenol/chloroform solutions at 4 °C in amber-colored reagent bottle. Always make sure to wear gloves while handling organic solvents.
4. The presence of isopropanol or ethanol in dried DNA samples can be identified via its smell. Proceed with dissolving pellet only when the alcohol smell is unobtrusive.

5. The ratio of $A_{260}/A_{280}$ between 1.7 to 2.0 is generally considered as pure DNA. Typically, ratio lower than 1.7 indicates the presence of contaminants in DNA, while an increase of this ratio suggests RNA contamination in the DNA sample.
6. Prior to oligonucleotide synthesis, the cross-reactivity of selected primers should be checked by searching oligonucleotide sequences against gene databases for available homology (NCBI-BLAST). This step will increase the specificity of oligonucleotides toward the target organism. For closely related fungi, the primer specificities must be checked experimentally.
7. DNA extracted from all fungi should be first tested using positive control primers that amplify fungal DNA to prove that the extracted DNA is amplifiable.
8. PCR diagnosis assay being reasonably simple and robust can be transferred between laboratories. However, PCR diagnostic methods require cautious laboratory practices to avoid contaminations. To reduce the risk of contamination in PCR reactions, prepare 0.2 mL aliquots of autoclaved $ddH_2O$ in sterile microtubes and store at −20 °C. After PCR reaction, discard the used aliquot.
9. Make sure to completely thaw PCR components on ice. To ensure homogeneity of PCR components, thoroughly mix them by gentle tapping before adding to the reaction tubes. After preparing PCR reactions, briefly centrifuge PCR tubes for 10–15 s to settle reaction mixture to the bottom of the PCR tube.
10. Ethidium bromide is a known mutagen and carcinogen. Always wear gloves while handling and take appropriate care to avoid skin contact.
11. The concentration of the reagents used in PCR reaction influences the specificity of PCR assay. In case of multiple and nonspecific DNA bands, optimize PCR amplification conditions by testing varying concentration of PCR components and conditions.
12. The detection of specific PCR product amplified from a unique gene present in one or few cells requires a higher number of cycles. This reduces the sensitivity of the reaction and requires optimization of the number of PCR cycles to avoid nonspecific DNA bands.
13. Contamination is the most common problem in the PCR seed assay leading to false-positive results. Various sources of contamination in PCR reaction include aerosols, hair, skin, gloves, contaminated reagents, or the presence of other organisms such as saprophytes and pathogens in the seed. Further, the

presence of PCR inhibitors in seeds including phenolic compounds, polysaccharides, proteins, and enzymes can reduce or inhibit the amplification reaction. Furthermore, the PCR-based detection assay can amplify DNA from dead mycelium, as the technique cannot differentiate between DNA from viable and nonviable microorganisms. Thus, while estimating the disease potential, risks may be overestimated. To prevent amplification from nonviable cells and interferences from PCR inhibitory substances present in seeds, only the mycelium growing on the seeds should be used for DNA extraction and only the fungal DNA should be used for PCR.

14. For the quantification of target DNA, real-time PCR can be used [14]. However, the high running costs and reproducibility issues related to real-time PCRs restrict their use for plant disease diagnosis. In some cases, PCR using a single pair of primer does not always give a sufficiently specific or sensitive test results. This issue can be solved by using nested PCR approach, where the product of first PCR is subjected to a second round of PCR using a new set of primers which are complementary to the amplicon generated [15].

## References


1. Long DH, Correll JC, Lee FN, TeBeest DO (2001) Rice blast epidemics initiated by infested rice grain on the soil surface. Plant Dis 85:612–616
2. Thinlay X, Finckh MR, Bordeos AC, Ziegler RS (2000) Effects and possible causes of an unprecedented rice blast epidemic on the traditional farming system of Bhutan. Agric Ecosyst Environ 78:237–248
3. Martin RR, Jamer D, Levesque CA (2000) Impacts of molecular diagnostic technologies on plant disease management. Annu Rev Phytopathol 38:207–239
4. Neergaard P, Lambat AK, Mathur SB (1970) Seed health testing of rice III: testing procedure for detection of *Pyricularia oryzae* cav. Proc Int Seed Test Assoc 35:165–168
5. Uddin W, Viji G, Schumann GL, Boyd SH (2003) Detection of *Pyricularia grisea* causing gray leaf spot of perennial ryegrass turf by a rapid immuno-recognition assay. Plant Dis 87:772–778
6. Mullis K (1987) Process for amplifying nucleic acid sequences. United States Patent No 4,683,202. http://patft.uspto.gov/netahtm/PTO/srchnum.htm
7. Saiki RK, Gelfand DH, Stoffel S, Scharf SJ, Higuchi R, Horn GT, Mullis KB, Erlich HA (1988) Primer-directed enzymatic amplification of DNA with a thermostable DNA polymerase. Science 239:487–491
8. Chadha S, Gopalakrishna T (2006) Detection of *Magnaporthe grisea* in infested rice seeds using polymerase chain reaction. J Appl Microbiol 100:1147–1153
9. Konstantinova P, Bonants PJM, Pelzer VG, Zouwen PVD, Bulk RVD (2002) Development of specific primers for detection and identification of *Alternaria* spp. in carrot material by PCR and comparison with blotter and plating assay. Mycol Res 106:23–33
10. Ma Z, Michailides T (2007) Approaches for eliminating PCR inhibitors and designing PCR primers for the detection of phytopathogenic fungi. Crop Prot 26(2):145–161
11. Untergasser A, Cutcutache I, Koressaar T, Ye J, Faircloth BC, Remm M, Rozen SG (2012) Primer3 – new capabilities and interfaces. Nucleic Acids Res 40(15):e115
12. Kibbe WA (2007) OligoCalc: an online oligonucleotide properties calculator. Nucleic Acids Res 35:W43–W46
13. White TJ, Bruns T, Lee S, Taylor J (1990) Amplification and direct sequencing of fungal ribosomal RNA genes for phylogenetics. In: Innis MA, Gelfand DH, Sninsky JJ, White TJ (eds) PCR protocols: a guide to method and

applications. Academic, San Diego, CA, pp 315–322

14. Sun G, Liu J, Li G, Zhang X, Chen T, Chen J, Zhang H, Wang D, Sun F, Pan H (2015) Quick and accurate detection and quantification of *Magnaporthe oryzae* in rice using real-time quantitative polymerase chain reaction. Plant Dis 99:219–224
15. Foster SJ, Ashby AM, Fitt BDL (2002) Improved PCR based assays for pre-symptomatic diagnosis of light leaf spot and determination of mating type of *Pyrenopeziza brassicae* on winter oilseed rape. Eur J Plant Pathol 108:379–383

# Chapter 16

# Isolation and Functional Analysis of Effector Proteins of *Magnaporthe oryzae*

**Meng-Di Dai, Yu Li, Li-Xiao Sun, Fu-Cheng Lin, and Xiao-Hong Liu**

## Abstract

In nature, plants have evolved a myriad of preformed and induced defenses to protect themselves from microbes. Upon microbial infection, the recognition of the microbe-associated molecular patterns (MAMPs) by the pattern recognition receptors (PRRs) triggers the first stage of defense response (Dodds and Rathjen, Nat Rev Genet 11:539–548, 2010). However, in order to develop microbial delivery, effectors target PRRs for deregulating immune responses and facilitating host colonization (Thomma et al., Plant Cell 23:4–15, 2011). Here, we contribute a protocol for the screening system of *Magnaporthe oryzae* effectors and construct a fluorescent system to trace secretory proteins in the sheath infection samples. Using the tobacco rattle virus (TRV) system, the proteins including LysM, Chitin, Cutinase, and CFEM domains were selected and divided into two kinds according to the results of cell death induced or inhibited test in *Nicotiana benthamiana*. Then, candidate effectors can be deleted or overexpressed in *M. oryzae*. The barley or rice infection with *M. oryzae*, rice leaf sheath inoculation, and subcellular localization during infection can be performed to explore the functions of these effectors.

**Key words** Effectors, *Magnaporthe oryzae*, Screening system, Subcellular localization, Rice leaf sheath inoculation assays

## 1 Introduction

When pathogens infect plants, they must block the host defenses and sustain the host processes requiring for their feeding and growth. The first line of the plant immune system involves basal defense responses that are triggered by the pathogen-associated molecular patterns (PAMPS). PAMPS-triggered immunity is a plant defense mechanism including ROS burst and the secretion of antimicrobial compounds, hydrolytic enzymes (proteinases, chitinases, and glucanases) that can inhibit hydrolytic enzymes of pathogen, and the hydrolytic enzymes are damaging to plants. The effectors secreted by pathogens are generally small unique proteins, and many of them function to defeat PTI [3]. In *M. oryzae*, the

Stefan Jacob (ed.), *Magnaporthe oryzae: Methods and Protocols*, Methods in Molecular Biology, vol. 2356,
https://doi.org/10.1007/978-1-0716-1613-0_16, 

Avrpiz-t functioned to inhibit pathogen-associated molecular pattern (PAMPS)-triggered immunity (PTI). In vitro and in vivo, Avrpiz-t interacted with APIP6 and APIP10, respectively. The ubiquitination assay demonstrated that Avrpiz-t suppressed APIP6 and APIP10-mediated ubiquitination, and in return, APIP6 and APIP10 ubiquitinated Avrpiz-t [4, 5]. Plants have developed a second line of defense, effector-triggered immunity (ETI), which involves the detection of specific avirulence effectors (AVR effectors) in plants. In many cases, it was indirectly inferred from the fact that the effectors matched by the resistance (R) proteins in plants were recognized as avirulence (Avr) proteins. The existence of matching Avr-R protein pairs resulted in a typical hypersensitive response (HR) in hosts, which can lead to cell death. Therefore, we combined the bioinformatics analysis with the tobacco rattle virus (TRV) system in *Nicotiana benthamiana* to isolate the effectors in *M. oryzae*.

Studies on the secretory structure of *M. oryzae* have been well established. There are two distinct types based on the localization of host cells in *M. oryzae* effectors. Type 1 are cytoplasmic effectors, which were secreted into plant cells, such as Pwl1, Pwl2, Avr-Pita, and Avr-Piz-t. They are preferentially accumulated in the biotrophic interfacial complex (BIC), a plant-derived interfacial structure, developed at the tip of primary invasive hyphae (IH). Such effectors are left behind when the primary IH differentiated into the bulbous pseudohyphae and continued to invade the neighboring plant cells, while the BIC remained at the same position and delivered newly synthesized effectors [6]. The efficient secretion of cytoplasmic effectors required the exocyst complex and t-SNARE [7]. Avr-Pita might be a zinc metalloprotease, which bind with Pi-ta protein in rice to initiate the defense response [8]. Pwl2 was a member of a dynamic, rapidly evolving gene family which escaped itself from infecting a second host grass, weeping lovegrass [9]. Type 2 are apoplastic effectors, which are retained in the extra-invasive hyphal membrane (EIHM) compartment including Bas4, Avr1-CO39, and Slp1. EIHM is plant-derived and surrounding the IH. By constructing the Bas4-GFP fusion protein, GFP signals were found appearing to outline the IH [10]. Avr-CO39 was a host specificity gene which was avirulent to rice cultivar CO39 because it could recognize the resistance gene Pi-CO39 in rice to trigger a hypersensitive response [11]. Slp1 contained LysM domain that competed with CEBIP in rice to bind with chitin oligosaccharides leading to a significant effect on virulence decrease [12]. Through establishing the fluorescent system of the *M. oryzae* effectors, we can distinguish the cytoplasmic effectors from apoplastic effectors for further research activities.

## 2 Materials

### 2.1 Organisms

1. The pathogen *M. oryzae* strain 70-15, cultured on complete medium (CM) at 25 °C (16:8 h, day/night) for 8–12 days.
2. The *Escherichia coli* strain DH5α contributes to extract the plasmids.
3. Rice cultivar CO39 (*Oryza sativa* L.), wheat cultivar (*Hordeum vulgare* L.), and tobacco (*Nicotiana benthamiana*) used for pathogenicity assays are grown in the greenhouse at 25 °C (16:8 h, day/night).

### 2.2 Antibiotics

Prepare stock solutions in $ddH_2O$ and filter with 0.22μm filter (Millipore, Germany): 100 mg/mL ampicillin, 400 mg/mL cefotaxime, 60 mg/mL streptomycin, 50 mg/mL hygromycin B, 50 mg/mL kanamycin. Solve 50 mg/mL sulfonylurea in DMSO (*see* **Note 1**).

### 2.3 Identification of Putative Proteins and Bioinformatics Analysis

1. PFAM, BLASTP in rice blast fungus genome-wide search database.
2. Rice blast fungus database (http://fungidb.org/fungidb/).
3. SMART (http://smart.embl-heidelberg.de/smart/change_mode.pl).
4. TargetP (http://www.cbs.dtu.dk/services/targetp/).

### 2.4 Screening Putative Effectors

1. Lysogenic broth medium (agar). 10 g/L tryptone, 5 g/L yeast extract, 10 g/L NaCl, 15 g/L agar, pH 7.5.
2. Lysogenic broth medium (liquid). 10 g/L tryptone, 5 g/L yeast extract, 10 g/L NaCl, pH 7.5.
3. *Agrobacterium tumefaciens* strain GV3101.
4. 0.2 M acetyleugenone (AS) (dissolve in DMSO).
5. 0.5 M MES. Dissolve 4.88 g MES in 50 mL of $ddH_2O$, pH 5.6.
6. Infiltrating buffer. 10 mM $MgCl_2$, 10 mM MES, 150 nM acetyleugenone.
7. *Nicotiana benthamiana* leaves (3 weeks old).
8. Tobacco rattle virus (TRV), which is a kind of widely used virus vector [13].
9. Primers for amplifying the effectors. Design the synthetic primer of each effector according to the specific effector sequence.
10. ClonExpress II One Step Cloning Kit (Vazyme, China).
11. AxyPrep-96 (Axygen, America).

### 2.5 Generation of the Null Mutants and Construction of the Fluorescent Vectors

1. Complete medium (agar). 10 g/L glucose, 2 g/L peptone, 1 g/L yeast extract, 1 g/L casamino acids, 0.1% (v/v) trace elements(2.2 g $ZnSO_4 \cdot 7H_2O$, 1.1 g $H_3BO_3$, 0.5 g $MnCl_2 \cdot 4H_2O$, 0.5 g $FeSO_4 \cdot 7H_2O$, 0.17 g $CoCl_2 \cdot 6H_2O$, 0.16 g $CuSO_4 \cdot 5H_2O$, 0.15 g $Na_2MoO_4 \cdot 5H_2O$, 5 g $Na_4EDTA$; dissolve in 100 mL of double distilled water), 0.1% (v/v) vitamin supplement (0.01 g biotin, 0.01 g pyridoxine, 0.01 g thiamine, 0.01 g riboflavin, 0.01 g PABA (p-aminobenzoic acid), and 0.01 g nicotinic acid; dissolve in 100 mL of double distilled water), 6 g/L $NaNO_3$, 0.5 g/L KCl, 0.5 g/L $MgSO_4$, 1.5 g/L $KH_2PO_4$, 15 g/L agar, pH 6.5.
2. *Agrobacterium tumefaciens*-induced medium (AIM): 0.8 mL 1.25 M potassium phosphate buffer pH 4.8, 20 mL MN-buffer (30 g/L $MgSO_4 \cdot 7H_2O$, 15 g/L NaCl), 1 mL of 1% $CaCl_2 \cdot 2H_2O$ (w/v), 10 mL of 0.01% $FeSO_4$ (w/v), 5 mL of AIM-salts (100 mg/L $ZnSO_4 \cdot 7H_2O$, 100 mg/L $CuSO_4 \cdot 5H_2O$, 100 mg/L $H_3BO_3$, 100 mg/L $Na_2MoO_4 \cdot 2H_2O$), 2.5 mL of 20% $NH_4NO_3$ (w/v), 10 mL of 50% glycerol (v/v), 40 mL of 1 M MES (pH 5.5), 10 mL (liquid medium) or 5 mL (solid medium) of 10 mL 20% glucose (w/v), 200μM acetosyringone. Add distilled water to 1 L and autoclave for 20 min. For solid medium add 1.5% agar before autoclaving.
3. Defined complex medium (DCM) (agar): 10 g/L glucose, 2 g/L asparagine, 1.7 g/L yeast nitrogen base without amino acids (Difco), 1 g/L $NH_4NO_3$, 15 g/L agar, pH 6.0.
4. Lysogenic broth medium (*see* Subheading 2.4).
5. The wildtype strain genomic DNA.
6. Primers of Target-upF: CCGGGGATCCTCTAGAGATG CATTCTCAATTACTTGGAAG.

   Target-upR: GATGTGGGGCACTGTGGCGTTGG CACTGTCTGCGGGCCTGTCCGAG.

   Target-downF: ATTGCATGCTCTCACGATTCA CAATCTGGTGTTTGGAATGGGTTGGTTTGCTGCA.

   Target-downR: GGCCAGTGCCAAGCTTTGC GATGTTGGCGAATAATCTTC (*see* **Note 2**).

   SUR-F: GTGCCAACGCCACAGTGCCCCAC.

   SUR-R: GATTGTGAATCGTGAGAGCATGCAATTC.

   Long-F: GTGTGCAGATGAAAGACTG.

   Long-R: GTTCTGCCCAAGCATGAGC.

   Short-F: CTTCTTCCTCACGGCCAACAACGG.

   Short-R: TCGGCAGGGTCCTCTAGGCATTCG.

   Tubulin-F: CCATCCCGAGCTTGTTGATA.

   Tubulin-R: GTAGTTCAGGTCACCGTATGAG (*see* **Note 3**).

   SUR-qF: CAAGGAGTGGAAGGAGAAGTG.

   SUR-qR: CCGGTTGTGATGTAGGTCATG.

Tubulin-qF: ACAACTTCGTCTTCGGTCAG.
Tubulin-qR: GTGATCTGGAAACCCTGGAG (*see* **Note 4**).

7. The skeleton vector of pKO1B, which contains the promoter of histone H3 in *M. oryzae* and the eGFP gene [14].
8. *Agrobacterium tumefaciens* strain AGL1.
9. The *Magnaporthe oryzae* wildtype strains: 70-15, Guy11.
10. ClonExpress MultiS One Step Cloning Kit (Vazyme, China).
11. AxyPrep-96.
12. Nikon fluorescence microscopy.
13. NANO drop.
14. Mastercycler ep realplex Thermal Cycler (Eppendorf, Hauppauge, NY).
15. TaKaRa TB Green™ Premix Ex Taq™ II (Tli RNaseH Plus) (Takara Bio Inc., Shiga, Japan).
16. Plasmid pCB1532 (fungal genetics stock center, FGN44:52-53).

### 2.6 Barley or Rice Infection with M. oryzae

1. Barley leaves (8 days old).
2. Glass basin.
3. Rice plants (14 days old).
4. Strains of wildtype and mutants (10 days old).
5. 0.4% gelatin solution. Mix 0.4 g of gelatin in 100 mL of double distilled water.
6. Sprayer.
7. 0.5 cm (diameter) of puncher.

### 2.7 Observation of Effector Localization

1. Complete medium (agar) (*see* Subheading 2.5).
2. *Agrobacterium tumefaciens*-induced medium (*see* Subheading 2.5).
3. Lysogenic broth medium (*see* Subheading 2.4).
4. The *Magnaporthe oryzae* wildtype strain 70-15.
5. Genomic DNA of wildtype.
6. PKD1-GFP vector with the hygromycin B-resistant gene [14].
7. AxyPrep-96.
8. *Agrobacterium tumefaciens* strain AGL1.
9. ClonExpress II One Step Cloning Kit.
10. Nikon fluorescence microscopy.
11. Primers of Target-effector-GFP-F:

ATCAATCACAATGGCCGGATCCATGAAGCAAAG AATCACAAG.

Target-effector-GFP-R:

CCTTGCTCACCATCCCGGGTGTGCACTGACCAAA CTGAGAC (*see* **Note 5**).

12. Rice leaf sheaths (3 weeks old).
13. 1 mL syringe.
14. Glass basin.

## 3 Methods

### 3.1 Identification of Putative Proteins

Perform Pfam domain analysis of the *M. oryzae* proteome using the integrated module of Pfam domain search in the CLC Genomics Workbench (QIAGEN) with default parameters. Use BlastP analysis to establish sequence relationships, with an e-value of 0.001 or lower defined as statistically significant. According to the characteristic of each domain, select the proteins containing LysM, Chitin, Cutinase, and CFEM. Obtain the genome and amino acid sequence from the rice blast fungus database.

### 3.2 Bioinformatics Analysis

Perform SMART and TargetP to predict N-terminal signal peptides, domains, and subcellular localization of putative secretory proteins.

### 3.3 Screening Putative Effectors Using the Transient Expression System of Nicotiana benthamiana

#### 3.3.1 Construction of the Transient Expression Vectors

1. Amplify the targeted genes from genomic DNA by primers of Target-effector-F/R (from start to stop codon).
2. Join the fragments and the linearized TRV vector together to construct transient expression vectors following the manufacture's protocol of ClonExpress II One Step Cloning Kit.
3. Extract the plasmids following the manufacturer's protocols of AxyPrep-96.

#### 3.3.2 The Infiltration Assay of N. benthamiana Leaves

1. Transform the plasmids into *A. tumefaciens* strain GV3101 and freeze in liquid nitrogen.
2. Defrost in 37 °C water bath for 5 min.
3. Add 800μL of lysogenic broth liquid medium. Shake at 28 °C, 100 rpm for 2–3 h.
4. Centrifuge the bacterial solution at 7965 × *g* for 1 min, discard part of the supernatant, and coat on the lysogenic broth medium containing 50μg/mL kanamycin.
5. Grow in the dark for 2 days at 28 °C.

6. Pick one colony into 5 mL of lysogenic broth medium containing 50μg/mL kanamycin. Shake at 28 °C, 200 rpm overnight, and store 1500μL of the solution in 20–30% glycerol.
7. Activate 200μL of *A. tumefaciens* strains by 5 mL of liquid lysogenic broth medium containing 50μg/mL kanamycin at 28 °C, 200 rpm for 12–16 h.
8. Add 100μL of activated bacterial solution to the lysogenic broth medium, containing 20μM acetosyringone and 10 mM MES, for 16–20 h.
9. Centrifuge the bacterial solution at 6372 × *g* at room temperature to acquire the bacterial solution precipitation.
10. Suspend the precipitation with infiltrating buffer and adjust $OD_{600}$ to 0.2. Then incubate for 3 h in the dark.
11. Inject the infiltrating bacterial solution into the 3-week-old *N. benthamiana* leaves, compared with the TRV-Bax and TRV-GFP as positive control and negative control, respectively (*see* **Note 6**).
12. Observe the phenotype after 4 days. Screen out the proteins, which could cause the obvious symptoms of the leaves for further study.

### 3.4 Generation of the Null Mutants

#### 3.4.1 Construction of the Knockout Vectors

1. Amplify the upstream and downstream flanking sequences from the wildtype genomic DNA using the primers Target-upF/Target-upR and Target-downF/Target-downR, respectively (*see* **Note 2**).
2. Clone a 3.0 kb *SUR* cassette from pCB1532 (fungal genetics stock center, FGN44:52-53) using the primers SUR-F and SUR-R.
3. Join the two flanking fragments, *SUR* cassette, and linearized pKO1B together to generate the knockout vector following the manufacturer's protocols of ClonExpress MultiS One Step Cloning Kit.
4. Culture *E. coli* DH5α in 5 mL of lysogenic broth medium containing 50μg/mL kanamycin at 37 °C, 200 rpm. Then store 1500μL in 20–30% glycerol.
5. Extract the plasmid following the manufacturer's protocols of AxyPrep-96.

#### 3.4.2 Generation of the Null Mutants

1. Transform the plasmids into the *A. tumefaciens* strain AGL1 as mentioned above (Subheading 3.3.2).
2. Harvest the conidia from the strain grown on CM for 10 days as described above. Wash the conidia with double distilled water and dilute to $1 \times 10^6$ conidia/mL.
3. Induce 400μL cells of *A. tumefaciens* by 5 mL of liquid *A. tumefaciens*-induced medium at 28 °C, 200 rpm for 5–6 h.

4. Co-culture 100μL of induced cells and 100μL of *M. oryzae* conidia on nitrocellulose membrane on solid *A. tumefaciens*-induced medium (AIM) in the dark at 22 °C for 2 days.
5. Cut the co-cultured nitrocellulose membrane into strips and put onto the selection DCM containing 100μg/mL sulfonylurea, 400μg/mL cefotaxime, and 60μg/mL streptomycin. Grow in the dark at 28 °C.
6. After 3–5 days, pick the putative mutants and transfer them onto DCM containing 100μg/mL sulfonylurea, 400μg/mL cefotaxime, and 60μg/mL streptomycin.
7. Observe the fluorescence of the putative mutants to identify insertion mutants by Nikon fluorescence microscopy.
8. Extract the genomic DNA for non-fluorescent strains.
9. PCR analysis: The first round PCR is called short-fragment PCR. Use the primers of Short-F/R and Tubulin-F/R to detect the targeted genes and *β-tubulin* gene. If there is only one band for *β-tubulin*, that means the target genes are knocked out. If there are two bands including the target genes and the *β-tubulin gene*, that means the transformant is in ectopic or the wildtype strain. Then the second-round PCR is called long-fragment PCR, in which the primers of Long-F/R are set to identify the product of one band in the first-round PCR. For null mutants, there is one band of 1.6–1.9 kb on the electrophoretic gel, and for ectopic transformants and wildtype strain, there is no band (*see* **Note 3**).
10. Quantitative PCR: Adjust the DNA concentration of transformants by NANO drop to 25 ng/μL. For real-time PCR, amplify 2μL DNA in a total volume of 20μL mixture containing 10μL of TB Green™ Premix Ex Taq™ II; 1μL of primers F/R, respectively (*see* **Note 4**); and 6μL of double distilled water on a Mastercycler ep realplex Thermal Cycler. If the copy number of the *SUR* gene is $1.0 \pm 0.2$ time, the mutants would be regarded as containing a single insertion of the *SUR* gene.
11. Inoculate the mutants on complete medium.

### 3.5 Barley or Rice Infection with M. oryzae

#### 3.5.1 Leaf-Cutting Inoculation on Barley

1. Select barley leaves and cut into 5 cm length.
2. Place the leaves face up in a glass basin covered with moist paper to keep the leaf alive.
3. Select three 0.5 cm (diameter) mycelial plugs from strains cultured for 8 days by punchers and put them onto the barley leaves. Perform two repeats for each strain.
4. Cover the glass basin with plastic film and place it in greenhouse at 25 °C (16:8 h, day/night).
5. After 4–5 days, observe the pathogenicity and record the data.

#### 3.5.2 Leaf-Cutting Inoculation on Rice

Prepare the third young leaf of the 14-day-old rice plants for the experiment. The method is the same as in barley leaf inoculation.

#### 3.5.3 Spray Inoculation on Rice

1. Prepare the 14-day-old rice plants as above.
2. Collect the conidia from the 10-day-old fungal strain, grown on CM. Wash the conidia with double distilled water and dilute to $1 \times 10^6$ conidia/mL.
3. Mix the conidial suspension and 0.4% gelatin solution of the same volume together to dilute the suspension to $5 \times 10^5$ conidia/mL.
4. Spray 2 mL of conidial suspension evenly on rice leaves with 0.2% gelatin as control. Perform two repeats for each strain.
5. Put the rice plants in the greenhouse under 25 °C in the dark for 2 days for colonization (*see* **Note 7**).
6. Transfer the rice plants to the greenhouse at 25 °C (16:8 h, day/night).
7. After 4–5 days, observe the pathogenicity and record the data.

### 3.6 Observation of the Effector Subcellular Localization

#### 3.6.1 Construction of the Fluorescent Vectors

1. Amplify the targeted genes from genomic DNA by primers of Target-effector-GFP-F/R (*see* **Note 5**).
2. Join the fragments and the linearized PKD1-GFP vector together to construct fluorescent vectors following the manufacture's protocol of ClonExpress II One Step Cloning Kit.
3. Extract the plasmids following the manufacturer's protocols of AxyPrep-96.

#### 3.6.2 Obtain the Fluorescent Strains

1–5: The same as in Subheadings 3.4.2, **steps 1–5**.

6. Cut the co-cultured nitrocellulose membrane into strips and put onto the CM medium containing 200μg/mL hygromycin B, 400μg/mL cefotaxime, and 60μg/mL streptomycin. Grow in the dark at 28 °C.
7. After 3–5 days, pick the putative mutants and transfer them onto the CM medium containing 200μg/mL hygromycin, 400μg/mL cefotaxime, and 60μg/mL streptomycin again.
8. Observe the fluorescence by Nikon fluorescence microscopy.

#### 3.6.3 Observation of the Localization in the Infection Stage

1. Transfer the fluorescent strains to a 9 cm CM plate for 10–12 days.
2. Wash the conidia with 5 mL of double distilled water and dilute to $1 \times 10^5$ conidia/mL.
3. Prepare the 3-week-old rice leaf sheaths.
4. Inject the conidial suspension into the rice leaf sheaths with a 1 mL syringe, compared with apoplastic effector *Bas4* and

cytoplasmic effector *PWL2* as control (*see* **Note 8**), and place the leaf sheaths face up in a glass basin covered with moist paper to keep the leaf sheaths alive.

5. Cover the glass basin with a plastic film and place it in greenhouse at 25 °C (16:8 h, day/night).
6. Select appropriate observation time to observe the location of the effector during the infection of rice leaf sheath cells (*see* **Note 9**). Excise infected tissue and put onto a glass slide and observe using a fluorescence microscope.

## 4 Notes

1. Dilute the initial concentration of the prepared antibiotics 1000 times before adding to the medium.
2. The target gene is MGG_05023 (1.25 kb); the Target-upF/Target-upR and Target-downF/Target-downR are primers that can amplify the upstream and downstream flanking sequence (about 1 kb) of target gene, respectively.
3. For PCR analysis, the primers of Short-F/R are limited in the targeted genes and primers Tubulin-F/R for the positive control *β-tubulin* gene. And for the long-fragment PCR, Long-F is designed outside the 5′–3′ flanking fragment of the target genes, and the other Long-R is limited in the *SUR* gene.
4. For quantitative PCR, use the wildtype strain *β-tubulin* (one copy in the genome) as control. The SUR-qF/SUR-qR and Tubulin-qF/Tubulin-qR are primers that can identify the copies of the SUR cassette gene and *β-tubulin* gene in transformants, respectively.
5. The Target-effector-GFP-F/R are primers that are designed to amplify the full-length target gene (from start to stop codon) which have homologous sequence with the linearized vector of PKD1-GFP.
6. Bax is one of the most important genes promoting apoptosis. The vector of TRV-Bax would cause the necrosis of *N. benthamiana* leaves and TRV-GFP would not. Observe the symptoms of leaves to judge the function of effector [15].
7. For spray inoculation on rice assay, we'd better keep the device sealed before we put the rice plants in the greenhouse.
8. Use the fluorescence localization of known apoplastic effector *Bas4* and cytoplasmic effector *PWL2* to determine the type of the unknown effector [10, 16].
9. Generally, when conidial suspension infiltrating into the leaf sheath of rice (24 h later), the observation time is appropriate.

## References


1. Dodds PN, Rathjen JP (2010) Plant immunity: towards an integrated view of plant-pathogen interactions. Nat Rev Genet 11:539–548
2. Thomma BP, Nürnberger T, Joosten MH (2011) Of PAMPs and effectors: the blurred PTI-ETI dichotomy. Plant Cell 23:4–15
3. Giraldo MC, Valent B (2013) Filamentous plant pathogen effectors in action. Nat Rev Microbiol 11:800–814
4. Park CH, Chen S, Shirsekar G, Zhou B, Khang CH, Songkumarn P, Afzal AJ, Ning Y, Wang R, Bellizzi M, Valent B, Wang GL (2012) The *Magnaporthe oryzae* effector AvrPiz-t targets the RING E3 ubiquitin ligase APIP6 to suppress pathogen-associated molecular pattern-triggered immunity in rice. Plant Cell 24:4748–4762
5. Park CH, Shirsekar G, Bellizzi M, Chen S, Songkumarn P, Xie X, Shi X, Ning Y, Zhou B, Suttiviriya P, Wang M, Umemura K, Wang GL (2016) The E3 Ligase APIP10 connects the effector AvrPiz-t to the NLR receptor Piz-t in rice. PLoS Pathog 12:e1005529
6. Zhang S, Xu JR (2014) Effectors and effector delivery in *Magnaporthe oryzae*. PLoS Pathog 10:e1003826
7. Giraldo MC, Dagdas YF, Gupta YK, Mentlak TA, Yi M, Martinez-Rocha AL, Saitoh H, Terauchi R, Talbot NJ, Valent B (2013) Two distinct secretion systems facilitate tissue invasion by the rice blast fungus *Magnaporthe oryzae*. Nat Commun 4:1996
8. Jia Y, McAdams SA, Bryan GT, Hershey HP, Valent B (2014) Direct interaction of resistance gene and avirulence gene products confers rice blast resistance. EMBO J 19:4004–4014
9. Kang S, Sweigard JA, Valent B (1995) The PWL host specificity gene family in the blast fungus *Magnaporthe grisea*. Mol Plant-Microbe Interact 8:939–948
10. Mosquera G, Giraldo MC, Khang CH, Coughlan S, Valent B (2009) Interaction transcriptome analysis identifies *Magnaporthe oryzae* BAS1-4 as biotrophy-associated secreted proteins in rice blast disease. Plant Cell 21:1273–1290
11. Peyyala R, Farman ML (2006) *Magnaporthe oryzae* isolates causing gray leaf spot of perennial ryegrass possess a functional copy of the *AVR1-CO39* avirulence gene. Mol Plant Pathol 7:157–165
12. Mentlak TA, Kombrink A, Shinya T, Ryder LS, Otomo I, Saitoh H, Terauchi R, Nishizawa Y, Shibuya N, Thomma BP, Talbot NJ (2012) Effector-mediated suppression of chitin-triggered immunity by *Magnaporthe oryzae* is necessary for rice blast disease. Plant Cell 24:322–335
13. Antoniw, J.F. and White, R.F. (1986) Changes with time in the distribution of virus and PR protein around single local lesions of TMV infected tobacco. Plant Mol Biol 6:145–149
14. Li H, Lu J, Liu X, Zhang L, Lin F (2012) Vectors building and usage for gene knockout, protein expression and fluorescent fusion protein in the rice blast fungus. J Agric Biotechnol 20:94–104
15. Lacomme C (1989) Bax-induced cell death in tobacco is similar to the hypersensitive response. Proc Natl Acad Sci U S A 96:7956–7961
16. Sweigard JA, Carroll AM, Kang S, Farrall L, Valent B (1995) Identification, cloning, and characterization of *PWL2*, a gene for host species specificity in the rice blast fungus. Plant Cell 7:1221–1233

# Chapter 17

# Analysis of Genetic Variations and Genomic Instabilities in *Magnaporthe oryzae*

**Sonia Chadha**

## Abstract

Retrotransposons are major components of the *Magnaporthe oryzae* genome; their high copy number and property of stable insertion in genome make them ideal tools to develop molecular markers. Retrotransposon-based marker techniques mainly rely on the amplification of DNA sequences present between the retrotransposon termini and some component of flanking genomic DNA. In this chapter, two marker systems known as inter-retrotransposon amplified polymorphism (IRAP) and retrotransposon-microsatellite amplified polymorphism (REMAP) are described for genetic diversity studies in *M. oryzae*. In the IRAP method, DNA profiles are generated using outward-facing primers from two nearby retrotransposons, while REMAP produces DNA profiles from genomic segments present in retrotransposons and microsatellite repeats. These marker techniques are simple, cost-effective, and easy to develop for polymorphism studies among *M. oryzae* isolates, races, or populations. In addition, the chapter also describes the utility of these retrotransposon-based DNA markers to study stress-induced genomic instabilities in *M. oryzae*.

**Key words** Genetic differentiation, IRAP, LTR-retrotransposons, Molecular markers, REMAP, Genotyping

## 1 Introduction

Rice blast caused by the ascomycete fungus *Magnaporthe oryzae* is mainly controlled by employing rice varieties resistant to blast pathogen. However, these varieties often lose resistance after a short period of time due to the highly variable nature of the pathogen population. Various factors influence the genetic diversity and genome stability of *M. oryzae* such as diverse hosts, agricultural systems, and environmental cues [1, 2]. Thus, understanding of the genetic diversity and dynamics of the pathogen population is essential for the blast disease management [3]. High genetic diversity and genomic instability of *M. oryzae* are often linked to the presence of a considerable portion of repetitive sequences in its genome, which includes diverse transposable elements. The class of long

Stefan Jacob (ed.), *Magnaporthe oryzae: Methods and Protocols*, Methods in Molecular Biology, vol. 2356,
https://doi.org/10.1007/978-1-0716-1613-0_17, 

terminal repeats (LTRs)-retrotransposons (LTR-RTNs) forms 3.8% of the *M. oryzae* genome [4], transposed via an RNA intermediate through a replicative mechanism, where the parental copy is preserved and new copies of retrotransposons move and insert at new genomic sites. This results in the high heterogeneity and rearrangements in *M. oryzae* genome, affecting the pathogenicity and host range genes, and also contributes to the fungal adaptability and survival under unfavorable conditions [2, 5].

Molecular markers have made it possible to analyze the structure and dynamics of the rice blast pathogen populations. Conventional methods such as amplified fragment length polymorphism (AFLP) and single-nucleotide polymorphism (SNP) detect single-nucleotide changes which have high reversion frequency [6]. Further, the polymorphic profiles of microsatellite or simple sequence repeat (SSR)-derived markers are based on the gain or loss of mostly less than 20 nucleotides [6]. On the contrary, retrotransposon DNA markers use PCR-based approaches to amplify a genomic segment present between the conserved LTR ends and the flanking genomic DNA. These marker assays detect large changes due to the integration of retrotransposons in different genomic locations. In this chapter, LTR-retrotransposon-based DNA marker techniques known as inter-retrotransposon amplified polymorphism (IRAP) and retrotransposon-microsatellite amplified polymorphism (REMAP) are presented. The IRAP assay uses one or two outward primers derived from either the 5′- or 3′-end of the LTR and detects insertional polymorphisms of retrotransposons. LTRs are direct repeats flanking the internal coding regions and are present at the 5′- and 3′-ends of the retrotransposon (Fig. 1).

LTR-retrotransposons are generally present in clusters and nest within each other. The outward-facing oligonucleotides derived from the LTR region amplify DNA fragments present between two nearby LTR-retrotransposons. DNA fingerprints generated using IRAP primers depend on the retrotransposon copy number, activity, and insertional patterns of retrotransposons of the given element. In REMAP assays, outward-facing LTR primer is used in combination with a simple sequence repeat (SSR) primer, which is designed for annealing at the 3′-end of SSRs. Consequently, a REMAP assay generates DNA bands by amplifying regions between LTR-retrotransposon and SSRs present in close proximity [7, 8]. The DNA profiles in IRAP and REMAP assays can also be generated from solo LTRs, which are formed due to intra-element recombination of a full-length element. Both IRAP and REMAP techniques amplify DNA fragments which can be easily resolved using agarose gel electrophoresis.

The IRAP and REMAP assays described in this chapter are based on *M. oryzae Ty3/Gypsy* LTR-retrotransposons Grasshopper [9], Maggy [10], and Pyret [11], which are 5.2, 5.4, and 7.2 kb in size, respectively. Their genomic structure consists of 5′ LTR,

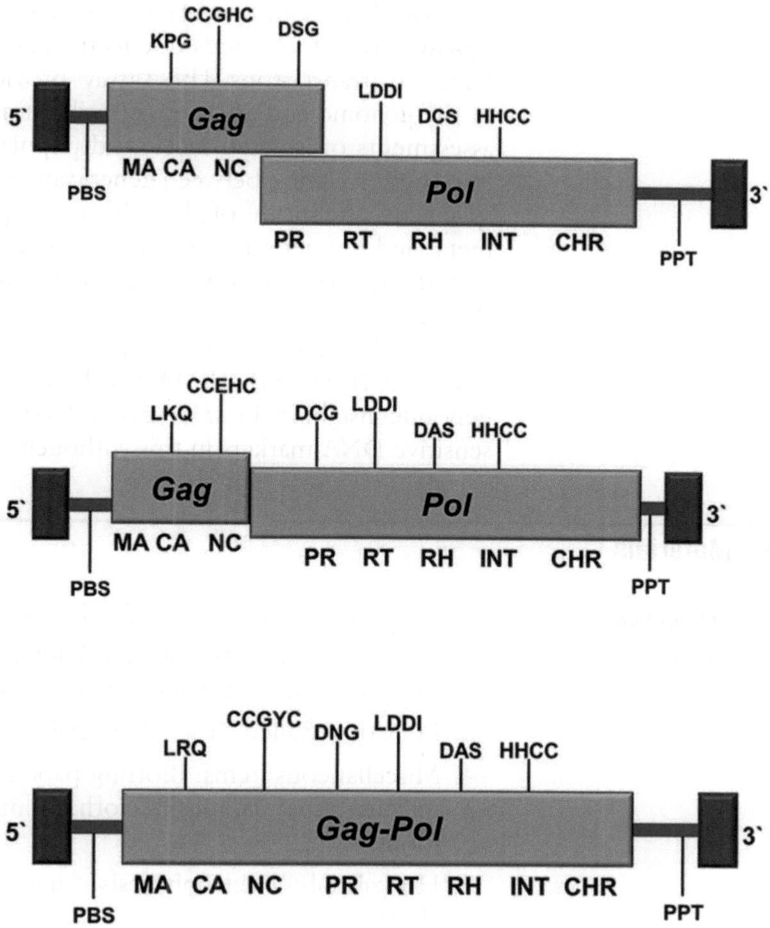


**Fig. 1** Gene structure of *M. oryzae Ty3/Gypsy* long terminal repeat (LTR) retrotransposons. Top, middle, and bottom panels represent gene structure and motifs of the *Magnaporthe* LTR-retrotransposons "Grasshopper," "Maggy," and "Pyret," respectively (figures not to scale). LTRs have been highlighted in black at termini and amino acid motifs noted with lines indicate the conserved residues in each protein domain. Abbreviations in the gene structure schemes mean the following: *ATF* aphid transmission factor, *CA* capsid, *CHR* chromodomain, *DU* or *DUT* dUTPase, *INT* integrase, *MA* matrix, *MOV* movement protein, *NC* nucleocapsid, *PBS* primer binding site, *PPT* polypurine tract, *PR* protease, *RH* RNaseH, *RT* reverse transcriptase, *SU* surface, *TAV* or *IBMP* transactivator/viroplasmin or inclusion body matrix protein, *TM* transmembrane, *VAP* virion-associated protein (Reproduced from Chadha and Sharma, 2020)

primer binding site (PBS), open reading frames (ORFs) for gag and pol genes, a polypurine tract (PPT), and 3′ LTR. Interestingly, the Pyret element has a single ORF gag-pol as compared to Maggy and Grasshopper that consists of two ORFs for gag-pol genes [12]. The long terminal repeat regions of Grasshopper, Maggy, and Pyret consist of 198, 253, and 475 nucleotides, respectively (Fig. 1).

Both IRAP and REMAP techniques generate a large number of polymorphic DNA bands due to the abundance and dispersion of LTR-retrotransposons. These assays provide the complete coverage of the genome and are of great utility for the genetic relationship assessments of isolates, races, and populations, providing insights into the correlation between genetic and pathotypic variability, host specificity, and means of plant disease propagation. Stress including metal and temperature induces transpositional activation of LTR-retrotransposons Grasshopper, Maggy, and Pyret, which results in the genomic rearrangements and instabilities in *M. oryzae* [2, 13]. These characteristics provide an excellent basis to apply advanced IRAP and REMAP techniques to study stress-induced genomic instability in *M. oryzae* and for the development of stress-sensitive DNA markers in this pathogen [2].

## 2 Materials

### 2.1 Equipment and Labware

1. Microcentrifuge tubes (1.5 mL), autoclave prior to use.
2. Clear microtips (10, 200, and 1000 μL), autoclave prior to use. Alternatively, use pre-sterilized filter tips.
3. PCR thin-walled clear tubes (200 μL), autoclave prior to use.
4. Miscellaneous items: Blotting paper, mortar and pestle, liquid nitrogen, spatula, muslin cloth, 50 mL Falcon tubes, and glass funnel.
5. Horizontal electrophoresis apparatus for agarose gel electrophoresis.
6. Gel combs with 24 or more well combs, 1 mm well spacing and 1 mm thickness.
7. Laminar flow bench or PCR workstation.
8. Variable volume micropipettes, dedicated for PCR and preferably autoclavable.
9. Power supply for electrophoresis.
10. Refrigerated centrifuge.
11. Refrigerated orbital shaker.
12. Thermal gradient cycler for 0.2 mL tubes or 96-well PCR plates (Eppendorf Mastercycler Gradient or equivalent).
13. Multimode reader with NanoQuant plate/NanoDrop or UV spectrophotometer for DNA quantification.
14. UV transilluminator for visualization of ethidium bromide-stained nucleic acids.
15. A digital gel imaging system for detection of ethidium bromide-stained nucleic acids.

16. Software such as Adobe Photoshop or GIMP—GNU Image Manipulation Program (freely available at https://www.gimp.org/)—for image analysis and manipulation.

### 2.2 Reagents

Prepare all solutions using Milli-Q/equivalent ultrapure water and molecular grade reagents.

1. Custom-synthesized LTR primers derived from the *Magnaporthe* LTR-retrotransposons "Grasshopper" [9], "Maggy" [10], and "Pyret" [11] are used [2, 12] in this study. Details of primers are given in Table 1. Simple sequence repeat primers used in REMAP assay [7, 8] are listed in Table 2. Dissolve lyophilized powder of oligonucleotides in sterile water to prepare stock solutions of 100 μM. Dilute to make working stock of 10 μM, prepare 100 μL aliquots, and store at −20 °C.
2. dNTP stock solution. Prepare 10 mM stock of dNTPs containing dATP, dCTP, dGTP, and dTTP in sterile water. Prepare aliquots of 100 μL and store at −20 °C.
3. 10× *Taq* DNA polymerase buffer. 100 mM Tris–HCl (pH 9.0), 500 mM KCl, 15 mM $MgCl_2$, and 0.1% gelatin.
4. DNA extraction buffer. Add 100 mM Tris–HCl (pH 8.5), 250 mM NaCl, and 25 mM EDTA in water and adjust pH to 8.0. Autoclave.

**Table 1**
**Details of LTR-retrotransposon-derived primers used in the study**

| LTR-retrotransposons | Primer | Primer sequence (5′–3′) |
|---|---|---|
| Grasshopper | GrhL2 | AGGAGGAGGGATGGGCAAGA |
| MAGGY | LTR1 | TGTCACAGACCTGAAGAACA |
| Pyret | PyR1 | CCCTTGTCCGTTTGAGATCA |

**Table 2**
**Selected microsatellite repeat primers for REMAP assay**

| Dinucleotide repeats | Dinucleotide repeat primers with 3′-anchored base |
|---|---|
| AG repeats | $(AG)_8C$, $(AG)_8G$ |
| GA repeats | $(GA)_8T$, $(GA)_8C$, $(GA)_8A$ |
| AC repeats | $(AC)_8G$ |
| CT repeats | $(CT)_8A/GC$, $(CT)_8A/GT$ |
| CA repeats | $(CA)_8A/GT$, $(CA)_8A/GC$, $(CA)_8A/GG$ |

5. Sodium acetate (3 M, pH 5.2). Dissolve 49.2 g of sodium acetate (molecular weight = 82.03) in $H_2O$. Use glacial acetic acid to adjust pH 5.2. Make final volume to 200 mL and autoclave.
6. Tris–EDTA (TE: 10 mM Tris, 1 mM EDTA) buffer, pH 8.0. Mix 1 mL of 1 M Tris–HCl buffer (pH 8.0), 200 μL of 0.5 M EDTA (pH 8.0), and 98.8 mL of $H_2O$. Autoclave and store at 4 °C.
7. RNase A (10 mg/mL) in 10 mM Tris–HCl (pH 8.0). Boil RNase A solution for 10 min and store as 200 μL aliquots at −20 °C.
8. DNA gel loading dye (6×). Mix 25 mg of bromophenol blue, 25 mg of xylene cyanol, 3.3 mL of glycerol, and 6.7 mL of sterile 10 mM Tris–HCl (pH 8.0). Aliquot and store at −20 °C for long-term use.
9. 10× Tris-borate-EDTA (TBE) buffer. Prepare 10× TBE buffer by dissolving 121.1 g Tris–HCl, 61.8 g boric acid, and 7.4 g EDTA (disodium salt) in $H_2O$, adjust pH to 8.0, and make final volume to 1 L. Autoclave and store for up to 6 months at room temperature.
10. Ethidium bromide stock (10 mg/mL) in sterile water.
11. Molecular weight DNA marker for electrophoresis 100–10,000 base range GeneRuler™ DNA Ladder Mix (Thermo Scientific) or similar. Dilute to 20 ng/μL for agarose gel electrophoresis.
12. Growth media. Prepare growth media in distilled water and autoclave at 121 °C for 15 min.
    - Potato dextrose broth (PDB). 20% potato infusion and 2% dextrose.
    - Potato dextrose agar (PDA). Add 2% (w/v) agar to PDB medium.
    - Complete media (CM). 3% yeast extract, 5% sucrose, and 3% casamino acids.

## 3 Methods

### 3.1 Genomic DNA Isolation

1. Grow *M. oryzae* isolates on PDA plates at 25 °C for 4–5 days. Inoculate a hyphal plug from the PDA culture plate in 100 mL PDB and grow at 25 °C for 5 days at 100 rpm.
2. Filter mycelium through a double layer of sterile muslin cloth. Remove traces of PDB by rinsing the mycelium with 200 mL of sterile water followed by drying between sterile blotting papers.

3. Grind fungal mass in liquid nitrogen to fine powder using mortar and pestle.
4. Immediately transfer mycelial powder (around 500 mg) to a 50 mL Falcon tube. Add 2.5 mL of DNA extraction buffer and 125 μL of 10% SDS. Mix and incubate at 65 °C for 30 min.
5. Add 850 μL of 5 M potassium acetate solution to sample tubes. Gently mix and incubate on ice for 30 min.
6. Add equal volume of chloroform/isoamylalcohol (24:1, v/v), mix, and centrifuge at 13,500 × *g* for 10 min (*see* **Note 1**).
7. Remove aqueous (upper) layer and add equal volume of ice-cold isopropanol. Mix and incubate at room temperature for 20 min. Centrifuge at 13,500 × *g* for 10 min. Discard supernatant and airdry the pellet in the laminar flow.
8. Dissolve DNA pellet in 1 mL of 10 mM TE buffer (pH 8.0) (*see* **Note 2**).
9. Add 10 μL of 10 mg/mL RNase A and incubate at 37 °C for 1 h.
10. Add an equal volume of a mixture of Tris–HCl-saturated phenol and chloroform (1:1, v/v). Mix and centrifuge for 5 min at 13,500 × *g* (*see* **Note 1**).
11. Remove aqueous (upper) layer and add chloroform/isoamylalcohol (24:1, v/v), mix, and centrifuge. Repeat this step.
12. Precipitate DNA sample by adding 1/10th volume of 3 M sodium acetate (pH 5.2) and two volumes of chilled ethanol to aqueous layer. Mix and incubate at −20 °C for 30 min. Centrifuge at 13,500 × *g* for 10 min at 4 °C.
13. Add 500 μL of 75% ethanol. Centrifuge at 13,500 × *g* for 5 min. Repeat this step.
14. Airdry the pellet in the laminar flow or PCR workstation and dissolve the pellet in 500 μL of 10 mM Tris–HCl buffer (pH 8.0) (*see* **Notes 2** and **3**).
15. Quantify genomic DNA samples by recording DNA absorbance at 260 and 280 nm using Nanodrop. The ratio of $A_{260}/A_{280}$ of 1.8:2.0 indicates purity of DNA (*see* **Note 4**).

### 3.2 Development of LTR-Retrotransposon Primers

1. Select *M. oryzae* LTR-retrotransposon elements from literature and obtain their DNA sequences from the GenBank database (NCBI) as fasta file. In this protocol, LTR primers derived from the *Magnaporthe* LTR-retrotransposons "Grasshopper" [9], "Maggy" [10], and "Pyret" [11] are used [2, 12]. Details of LTR primers are given in Table 1. Fasta DNA sequences of these elements, with accession numbers M77661.1 (Grasshopper), L35053.1 (Maggy), and AB062507.1 (Pyret), were obtained from the GenBank database (https://www.ncbi.nlm.nih.gov/genbank/).

2. Use DNA sequences of these elements (full gene sequence with 5′- and 3′-repeat regions) to construct the gene structure using the Gypsy Database (GyDB) of mobile genetic elements (http://gydb.org/index.php [14]) to predict LTR regions with confidence (Fig. 1).
3. Design outward primers to match LTR sequence near to either its 5′- or 3′-end (*see* **Note 5**). To design optimal PCR primers, online tool such as Primer3 (http://bioinfo.ut.ee/primer3/ [15]) can be used, which is also freely available for download.
4. Various parameters such as G+C content, primer melting temperature ($T_m$), product size range, and intra- and/or inter-complementarity of primers can be selected and runs of more than four bases are rejected. Properties of the designed primers can be analyzed using OligoCalc software (http://biotools.nubic.northwestern.edu/OligoCalc.html, [16]) (*see* **Note 6**).
5. Of the various primers designed by the software, shortlist selected primers. Prefer primers having proximity to the ends of the retrotransposon and with selectivity toward the retrotransposon. Avoid primers having internal secondary structure.
6. Custom synthesize primers commercially from available suppliers.
7. Prepare primer stock (100 μM) by dissolving oligonucleotides in sterile water. For routine use, make working primer stocks of 10 μM concentration. Store at −20 °C.

### 3.3 Inter-retrotransposon Amplified Polymorphism (IRAP)

#### 3.3.1 PCR Amplifications for Screening IRAP Primers

1. Outward primers derived from LTRs of "Grasshopper," "Maggy," and "Pyret" retrotransposons are used in this protocol. Primers are shortlisted based on the criteria mentioned in Subheading 3.2.
2. Occasionally, not all selected primers including those derived from retrotransposons work in the PCR. Perform primer screening to test LTR primers for amplification of DNA profiles using two geographically diverse isolates.
3. Set up all PCR experiments on ice in a clean laminar flow bench. Perform PCR reactions in 0.2 mL clear and thin-walled PCR tubes.
4. Perform IRAP amplifications in 25 μL volume containing 2.5 μL of 10× *Taq* DNA polymerase buffer, 50 ng of genomic template DNA, 0.6 μM of IRAP primer, 150 μM dNTPs (10 mM stock of dNTPs), and 0.5 U of *Taq* DNA polymerase. Add sterile PCR $H_2O$ to final volume of 25 μL. Include negative control (without DNA template) in all the PCR reactions (*see* **Notes 7–11**).

5. Perform PCR assay using the program: 95 °C, 4 min. 40 cycles of 94 °C, 30 s; 55 °C, 45 s; 72 °C, 2 min. 72 °C, 10 min (*see* **Notes 9** and **10**).
6. PCR tubes containing amplified products can be stored overnight at 4 °C, preferably without loading buffer.

#### *3.3.2 Agarose Gel Electrophoresis*

1. Prepare 2% (w/v) agarose gel by dissolving 4.0 g of agarose gel powder in 200 mL of 0.5× TBE buffer (pH 8.0) in a 500 mL flask. Dissolve and melt the agarose gel in a microwave oven (*see* **Note 12**).
2. Cool down agarose gel to around 60 °C and add 0.5 μg/mL (final concentration) ethidium bromide solution. Mix and cool to 50 °C (hand-warm) and slowly pour the agarose into the gel tray pre-sealed with thick tape or gasket. Allow gel to solidify at room temperature for at least 1 h. Fill the chamber with 0.5× TBE running buffer until the buffer reaches about 3–5 mm over the surface of the gel (*see* **Note 13**).
3. For resolving PCR samples, mix PCR-amplified products (20 μL) with 4 μL of 6× gel loading dye. Mix and briefly centrifuge for 5–15 s. Load the gels with a PCR sample (20 μL) and run the gel at 100 V. Use 100 bp DNA ladder as a molecular weight marker (*see* **Note 14**).
4. Visualize and photograph the agarose gel under ultraviolet illumination using gel documentation and image analysis systems.

#### *3.3.3 IRAP Assay for Genotyping*

1. From primer screening experiment (Subheading 3.3.1), select IRAP primers that give clear, distinct, polymorphic, and reproducible DNA profiles. Examples of IRAP primers are listed in Table 1.
2. For genotyping, use purified and quantified genomic DNA samples of *M. oryzae* isolates, races, or populations. Perform IRAP-PCR assay as described in Subheading 3.3.1.
3. Resolve IRAP bands using agarose gel electrophoresis as described in Subheading 3.3.2 (*see* **Note 15**).
4. Representative IRAP gel electrophoretic image is presented in Fig. 2.
5. Test the reproducibility of DNA profiles, and repeat PCR amplifications.

### 3.4 Retrotransposon-Microsatellite Amplified Polymorphism (REMAP)

1. To generate a REMAP profile, use the LTR1 primer (Table 1) derived from the "Maggy" retrotransposon in combination with ISSR primers (Table 2; *see* **Notes 15** and **16**).
2. The details of 11 microsatellite repeat primers that generate REMAP profiles with LTR1 primer are given in Table 2 (*see* **Note 16**).

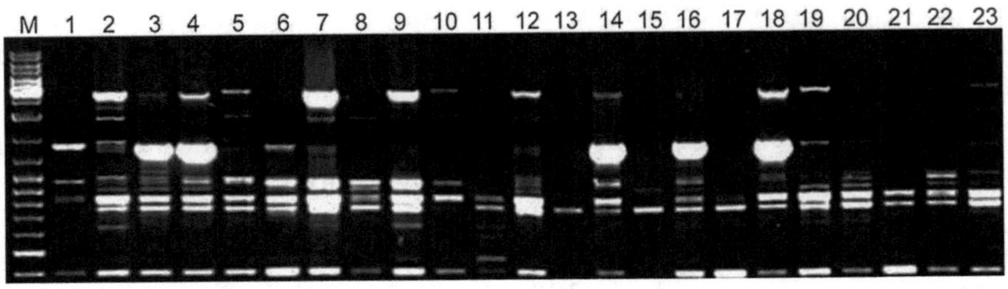


**Fig. 2** Genotypic profiles of *M. oryzae* isolates from diverse geographical regions of India and Japan using PyR1 outward primer derived from LTR-retrotransposon Pyret (Adapted from Chadha and Sharma, 2014)

3. Perform REMAP amplifications in 25 μL reaction volume containing 2.5 μL of 1× *Taq* polymerase buffer, 0.5 U *Taq* polymerase, 150 μM of each dNTP, 0.4 μM of LTR1, 0.6 μM of ISSR primer, and 50 ng of template DNA (*see* **Notes 7–11**). Add sterile water to a final volume of 25 μL.
4. Perform PCR using program: 92 °C, 5 min. 40 cycles of 92 °C, 45 s; 55 °C, 45 s; 72 °C, 1 min. 72 °C, 10 min (*see* **Note 9**).
5. Resolve PCR-amplified DNA using agarose gel electrophoresis as described in Subheading 3.3.2.
6. Test the reproducibility of DNA profiles, and repeat PCR amplifications with the selected primer pairs.

### 3.5 Data Analysis for IRAP and REMAP Assays

1. Consider each amplification product identifiable after electrophoresis as a DNA marker and scored across all samples. Score only reproducible and distinct bands (signals).
2. For scoring DNA bands, record the presence of a band in a given sample as 1 and the absence of a band as 0.
3. Interpolate the size of DNA bands using the relative migration distance of molecular weight DNA marker bands.
4. Analyze the genotyping data to calculate following parameters: the number of total bands (NTB), the number of polymorphic bands (NPB), percent polymorphism (PP), low frequency bands (LFB), bands per primer (BPP), and polymorphic bands per primer (PBP).

### 3.6 Detection of Genomic Instabilities in Stress-Treated M. oryzae Cultures

#### 3.6.1 Fungal Culture and Stress Treatment

1. Inoculate *M. oryzae* isolates on PDA plates at 25 °C for 5 days.
2. For DNA extraction, inoculate a 5 mm piece of agar with actively growing hyphae in 100 mL of CM media and incubate at 25 °C at 100 rpm.
3. After 4 days of growth, expose cultures to stress condition. This protocol is optimized for copper stress (0–2.5 mM) for 18 h and heat shock at 42 °C (1–3 h).

*3.6.2 Detection of Genomic Instabilities*

1. Isolate, purify, and quantify genomic DNA from stress-treated (test) and untreated (control) fungal cultures as described in Subheading 3.1.
2. To detect genomic changes in stress-treated samples as compared to the untreated (control), perform IRAP and REMAP assays using primers listed in Tables 1 and 2. Set up PCR reactions as described in Subheadings 3.3 and 3.4.
3. Resolve PCR-amplified bands of test and control samples using agarose gel electrophoresis as described in Subheading 3.3.2.

*3.6.3 Data Analysis*

1. Consider each reproducible amplified product across all samples (test and control) as a scorable DNA marker.
2. To determine the presence and absence of each band, make a binary matrix (1 for band presence and 0 for absence) for each sample. Also score changes of band intensities across all test samples as compared to control (*see* **Note 17**).
3. Calculate mutation rate (MR) of each treatment group using altered DNA bands. MR can be calculated as (Number of altered bands in test sample/total number of bands in control) × 100.
4. Calculate genomic template stability (GTS) as $100 - (100\ a/n)$ where, $a =$ number of altered bands in test samples and $n =$ number of bands in the control sample [17].
5. Analyze results by performing two-way analysis of variance (ANOVA) to confirm the statistical significance of the data.

## 4 Notes

1. Store chloroform/isoamylalcohol and phenol/chloroform solutions at 4 °C in amber-colored reagent bottle. Always make sure to wear gloves while handling organic solvents. Before performing the DNA purification step, check the compatibility of labware with chloroform and phenol.
2. The presence of isopropanol or ethanol in dried genomic DNA samples can be identified via its smell. Dissolve pellets only when the alcohol smell is unobtrusive.
3. Store DNA in 10 mM TE buffer (pH 8.0) at −20 °C. For routine use, prepare working stocks by diluting it to 100 ng/μL concentration in 10 mM Tris–HCl buffer (pH 8.0). The working stock can be stored at 4 °C for a few months.
4. DNA quality is important to obtain good and reproducible DNA profiles. The presence of PCR-interfering substances such as polysaccharides, pigments, excess salts, phenol, and

ethanol traces in DNA samples can reduce the efficiency of the PCR. DNA samples with poor quality generate short bands upon storage for a month or more.

5. The outward orientation of LTR primers allows the amplification of regions between LTR-retrotransposons and solo LTRs present in close proximity.
6. For each primer, calculate the optimal annealing temperature (Ta), where the PCR efficiency is maximal and only a specific product is amplified. Use temperature gradient option in thermal cycler to find the optimal annealing temperature for each primer.
7. To reduce the risk of contaminations in the PCR reactions, prepare 0.2 mL aliquots of autoclaved Milli-Q or ultrapure water in sterile microtubes and store at −20 °C. After PCR reaction, discard the used aliquot. Make sure to completely thaw PCR components on ice.
8. To ensure homogeneity of PCR components, thoroughly mix them by gentle tapping before adding to the reaction tubes.
9. To obtain clear and distinct bands, optimize the concentrations of PCR components including primers, dNTPs, DNA template, and $MgCl_2$. The efficiency of the PCR reaction also relies on the type of buffer-enzyme combination used. Most of the DNA polymerases are supplied commercially with their own recommended buffer. These thermostable polymerase buffers may also be usable with other polymerases. In addition, optimize conditions for the PCR program by varying denaturation time (30–60 s), annealing temperature (55–65 °C), annealing time (30–60 s), and the number of PCR cycles.
10. The concentration of the DNA template used in PCR depends on the source and the DNA amount may vary from 25 to 200 ng per reaction mix (25 μL). The amount of DNA used for PCR reactions depends on the concentration of PCR components, the number of target genes, and the quality of DNA. For detection of rare PCR products in the background of billion cells, a higher amount of starting DNA template and cycles is used. Since higher DNA concentrations result in smearing patterns, the amount of DNA needs to be optimized by testing the range of DNA concentrations in PCR reaction. Moreover, the number of cycles required to produce a sufficient amount of PCR products also depends on the DNA concentration. In general, for multi-locus profiling of fungal DNA, recommended DNA cycles are 35–40.
11. After preparing PCR reactions, briefly centrifuge PCR tubes for 10–15 s to settle reaction mixture to the bottom of the PCR tubes.

12. The casting of agarose gel is critical to the success of any gel-based genotyping assay. Melt and dissolve agarose gel completely in the buffer as small undissolved pieces may affect the quality of gel and resolution of DNA bands.
13. Always cool agarose gel evenly before casting by shaking it at 37 °C for around 30 min. Always wear gloves while handling ethidium bromide as it is a known mutagen. Handle carefully and take appropriate care to avoid skin contact.
14. Overloading of lanes may result in poor gel resolution. Electrophoresis may cause the gels to deteriorate after several hours; their temperature should not be allowed to exceed 30 °C, above which electrophoretic resolution will be impaired.
15. Generally, it takes around 1–2 months to develop a new LTR--retrotransposon-based marker for an organism in which retrotransposons have not been previously described. The development of these markers depends on the experience in primer design and sequence analysis and the pace of analysis. Once optimized primers and optimized PCR conditions are available, routine PCR reactions and analysis can be easily performed at a rapid pace.
16. The selection of ISSR primers for REMAP assay is mainly based on the following criteria: i) primers with an annealing temperature close to that of the LTR1 primer and ii) primers amplifying a pattern different from the pattern obtained when either primer was used alone.
17. A change in band intensity is defined as an increase or decrease of the intensity of the signal of a band by $\geq$ 50% in the stress DNA by direct visualization when compared to normal DNA [18].

## References


1. Thinlay X, Finckh MR, Bordeos AC, Ziegler RS (2000) Effects and possible causes of an unprecedented rice blast epidemic on the traditional farming system of Bhutan. Agric Ecosyst Environ 78:237–248
2. Chadha S, Sharma M (2014) Transposable elements as stress adaptive capacitors induce genomic instability in fungal pathogen *Magnaporthe oryzae*. Plos One 9(4):e94415
3. Zeigler RS, Leong SA, Teeng PS (1994) *Rice blast disease*. CAB International Wallingford in Association with the International Rice Research Institute, Los Baños
4. Dean RA, Talbot NJ, Ebbole DJ, Farman ML, Mitchell TK, Orbach MJ et al (2005) The genome sequence of the rice blast fungus *Magnaporthe grisea*. Nature 434:980–986
5. Kang S, Lebrun MH, Farall L, Valent B (2001) Gain of virulence caused by insertion of Pot3 transposon in *Magnaporthe grisea* avirulence gene. Proc Natl Acad Sci U S A 14:671–674
6. Kalendar R, Schulman AH (2006) IRAP and REMAP for retrotransposon-based genotyping and fingerprinting. Nat Protoc 1:2478–2484
7. Chadha S, Gopalakrishna T (2005) REMAP markers for genetic diversity assessment of the rice blast pathogen (*Magnaporthe grisea*). Genome 48:943–945
8. Chadha S, Gopalakrishna T (2007) Comparative assessment of REMAP and ISSR marker assays for genetic polymorphism studies in *Magnaporthe grisea*. Curr Sci 93:688–692
9. Dobinson KF, Harris RE, Hamer JE (1993) Grasshopper, a long terminal repeat (LTR)

retroelement in the phytopathogenic fungus *Magnaporthe grisea*. Mol Plant-Microbe Interact 6:114–126

10. Farman ML, Tosa Y, Nitta N, Leong SA (1996) *MAGGY*, a retrotransposon in the genome of the rice blast fungus, *Magnaporthe grisea*. Mol Gen Genet 251:665–674
11. Nakayashiki H, Matsuo H, Chuma I, Ikeda K, Betsuyaku S, Kusaba M, Tosa Y, Mayama S (2001) Pyret, a *Ty3/Gypsy* retrotransposon in *Magnaporthe grisea* contains an extra domain between the nucleocapsid and protease domains. Nucleic Acids Res 29:4106–4113
12. Chadha S, Sharma M (2020) Genetic differentiation and phylogenetic potential of *Ty3/Gypsy* LTR retrotransposon markers in soil and plant pathogenic fungi. J Basic Microbiol 2020:1–9. https://doi.org/10.1002/jobm.201900487
13. Ikeda K, Nakayashiki H, Takagi M, Tosa Y, Mayama S (2001) Heat shock, copper sulphate and oxidative stress activate the retrotransposon *MAGGY* resident in the plant pathogenic fungus *Magnaporthe grisea*. Mol Gen Genet 266:318–325
14. Llorens C, Futami R, Covelli L, Domínguez-Escribá L, Viu JM, Tamarit D et al (2011) The Gypsy Database (GyDB) of mobile genetic elements: release 2.0. Nucleic Acids Res 39: D70–D74
15. Untergasser A, Cutcutache I, Koressaar T, Ye J, Faircloth BC, Remm M, Rozen SG (2012) Primer3 – new capabilities and interfaces. Nucleic Acids Res 40(15):e115
16. Kibbe WA (2007) OligoCalc: an online oligonucleotide properties calculator. Nucleic Acids Res 35(Web Server issue):W43–W46
17. Liu W, Yang YS, Zhou Q, Xie L, Li P, Sun T (2007) Impact assessment of cadmium contamination on rice (*Oryza sativa L.*) seedlings at molecular and population levels using multiple biomarkers. Chemosphere 67:1155–1163
18. Maeda T, Jikko A, Hiranuma H, Fuchihata H (1999) Analysis of genomic instability in squamous cell carcinoma of the head and neck using the random amplified polymorphic DNA method. Cancer Lett 138:183–188

# Chapter 18

# Controllable Bypass Suppression in *Magnaporthe oryzae*

**Stefan Jacob and Katharina Bersching**

## Abstract

Evolutionary adaptation of living organisms is commonly thought to be the result of processes that have taken place over long periods of time. By contrast, we found that the filamentous rice blast fungus *Magnaporthe oryzae* rapidly suppresses the osmosensitive "loss of function" (lof) phenotype in knockout mutants of the high-osmolarity glycerol (HOG) pathway. That suppression occurs highly reproducibly after 4 weeks of continuous growth upon salt stress. Stable mutants reestablished in osmoregulation arise independently out of individual osmosensitive lof mutants of the HOG pathway. The major compatible solute produced upon salt stress by these suppressor strains was found to be glycerol, whereas it is arabitol in the wildtype strain. We aim to address the molecular or biochemical mechanisms behind this rapid suppression and characterize the associated factors and signaling pathways which enable or prevent suppression. Therefore, we present a protocol to generate these suppressor mutants in *M. oryzae* easily to study the molecular basis of evolutionary processes or even epigenetic modulation. This protocol may be applicable to many other fungi and will open a door for researchers worldwide since the HOG pathway is worked on intensively in many different model organisms.

**Key words** High-osmolarity glycerol (HOG) pathway, Adaptation, Bypass suppressor, Osmolyte production, Glycerol, Arabitol, *Magnaporthe oryzae*, Suppressor mutation, Rewiring, Polyol metabolism

## 1 Introduction

Genetic suppression is a term describing mutations which restore the original phenotype seen prior to the original background mutation. Historically, suppressors have proven to be extremely valuable since the early 1970s for determining the relationship between two gene products in vivo, even in the absence of cloning or sequence information [1–5]. Suppressor mutations are useful for identifying new genetic sites which affect a biological process of interest. They also provide evidence between functionally interacting molecules and intersecting biological pathways [2]. On one hand, mutations that restore a wildtype phenotype despite the continued presence of the original mutation are termed suppressors. On the other hand, modifiers that result in a more severe phenotype are termed

Stefan Jacob (ed.), *Magnaporthe oryzae: Methods and Protocols*, Methods in Molecular Biology, vol. 2356,
https://doi.org/10.1007/978-1-0716-1613-0_18, 

enhancers. The term suppressor was first introduced into *Magnaporthe* research by the identification of one of the most prominent bypass suppressors, the Mac1 phenotype (sum), involved in cAMP signaling [6]. The diversity of possible suppressor mechanisms can present a challenge in defining the relationship between a suppressor and the original gene. The simplest suppression mechanism is *intragenic suppression*, where a phenotype caused by a primary mutation is complemented by a second mutation in the same gene. *Informational suppressors* alter the passage of information from DNA to protein, in apparent violation of the genetic code. Furthermore, the class of *amount suppressors* consists of mutations that increase the amount of the original protein. Another mechanism for increasing the overall activity of a defective protein is to increase its specific activity. The respective mutations are called *activity suppressors.* A mutation that alters one step of the multistep pathway can often be suppressed by mutations in genes that affect other steps within that same pathway. This class of suppressors is often extremely informative, because in addition to identifying other components of the pathway of interest, the suppressors can also reorder the pathway [7]. A mutation that inactivates one pathway can often be suppressed by altering a second pathway. This *bypass suppressor* might affect the regulation of a pathway that has a related or overlapping function, or the suppressor could alter the specificity of a functionally unrelated pathway. The use of innovative suppressor hunts is important for deciphering biological pathways now and in the future. However, given the importance and the various possibilities of different types of suppression, knowledge of the molecular mechanisms behind and the molecular basis of what exactly facilitates, respectively, constrains suppressions are not well-documented to date. Therefore, a better understanding of the molecular mechanisms behind suppressors is required.

We describe a method in this protocol which explains how to generate suppressor mutants arising from "loss of function" (lof) mutants of the high-osmolarity glycerol (HOG) signaling pathway in *M. oryzae.* Inactivation of the components of the HOG pathway results in mutant strains, which are viable but impaired in osmoregulation [8]. After 4 weeks of cultivation upon high osmolarity, stable individuals with reestablished osmoregulation capacity arise independently from each of the lof mutants [9]. This phenomenon is highly reproducible and occurs only in osmosensitive lof mutants related to the HOG pathway and not in other osmosensitive *Magnaporthe* mutants. The major compatible solute produced by these suppressor strains to cope with high osmolarity is glycerol, whereas it is arabitol in the wildtype strain. Scientists will be able to follow up the question of how eukaryotic signaling pathways evolve to adapt toward changing environmental situations with the following protocol. This is one of the challenging questions in biology.

Adaptive evolution is a central biological process that underlies diverse phenomena, from the acquisition of antibiotic resistance to the evolution of niche specialization.

## 2 Materials

### 2.1 Generation of Suppressor Mutants (Suppressor Assay)

1. Fungal lof mutant strains of the HOG pathway (e.g., *Magnaporthe oryzae* Δ*Moypd1*, Δ*Mossk1*, Δ*Mossk2*, Δ*Mopbs2*, or Δ*Mohog1*) (*see* **Note 1**).
2. Petri dishes.
3. Complete medium (CM). To make 1 L, mix 10 g glucose, 1 g yeast extract, 2 g peptone, 1 g casamino acids, 50 mL nitrate salt solution (containing per liter $H_2O$: 120 g $NaNO_3$, 10.4 g KCl, 30.4 g $KH_2PO_4$, 10.4 g $MgSO_4{\cdot}7H_2O$), and 1 mL of a trace element solution (containing per liter $H_2O$: 22 g $ZnSO_4{\cdot}7H_2O$, 11 g $H_3BO_3$, 5 g $MnCl_2{\cdot}4H_2O$, 5 g $FeSO_4{\cdot}7H_2O$, 1.7 g $CoCl_2{\cdot}6H_2O$, 1.6 g $CuSO_4{\cdot}5H_2O$, 1.5 g $Na_2MoO_4{\cdot}2H_2O$, 50 g $Na_2EDTA$, pH 6.5 adjusted by 1 M KOH).
4. Osmotic stress medium. Mix the CM with stress-inducing agents, for example, KCl, NaCl, or sorbitol (*see* **Note 3**).
5. Osmotic stress stock solutions. Exemplarily, solve 2.5 M KCl in $H_2O$.
6. Autoclave.
7. Growth chamber or incubator with light.

All chemicals used were p.a. quality unless otherwise stated. All preparations should be carried out with deionized water ($H_2O$).

### 2.2 Verification of the Suppressor Strains

1. DNeasy® Plant Mini Kit (Qiagen).
2. Individual restriction enzymes (depends on which gene knockout is to be validated).
3. Reagents and equipment for standard Southern analysis.
4. Further HPAEC-PAD analysis is recommended for a reliable and comprehensive testing of the suppressor strains generated (*see* Chapter 4). Consequently, reagents and equipment for HPAEC-PAD analysis are needed.

## 3 Methods

### 3.1 Generation of Suppressor Mutants (Suppressor Assay)

1. Use the lof mutant strains with an inactivated HOG pathway (e.g., Δ*Moypd1*, Δ*Mossk1*, Δ*Mossk2*, Δ*Mopbs2*, and Δ*Mohog1*).
2. Begin growing each fungus from the filter stocks (*see* **Note 2**) and incubate them about 8–10 days at 26 °C on solid CM, light/dark 16/8 h (in petri dishes).

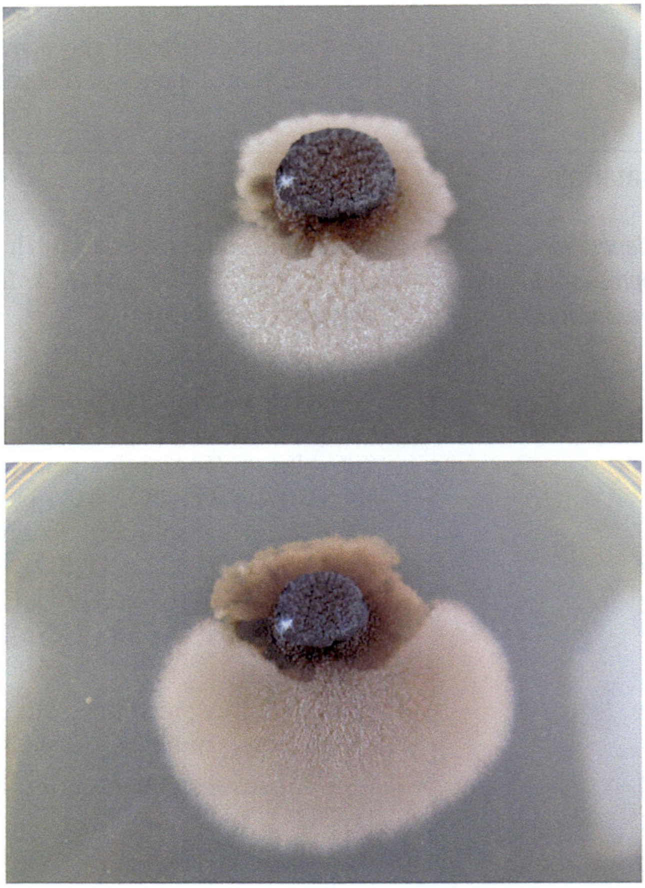

**Fig. 1** Example of *Magnaporthe oryzae* "loss of function" (lof) strain *ΔMohog1* and *ΔMohog1(suppressor)* after 5 weeks of cultivation upon 1 M KCl salt stress. The lof strain *ΔMohog1* (dark brown) was grown on CM including 1 M KCl at 26 °C with light/dark 16/8 h. The *ΔMohog1* mutant was found to be highly sensitive toward 1 M KCl, whereas the outgrowing *ΔMohog1(suppressor)* strain (bright) was able to grow much faster

3. Transfer a small plug from each plate to new CM plates and incubate for 8–10 days before starting the suppressor assay.
4. Transfer a small plug from a plate that is growing in the incubator onto osmotic stress medium (e.g., solid CM including 1 M KCl) (*see* **Note 3**).
5. Cultivate the fungal cultures at 26 °C with a light/dark rhythm of 16/8 h.
6. After about 4 weeks, small mycelium filaments begin to grow out of the original cultures (Fig. 1; *see* **Note 4**). These mycelium filaments grow much faster as compared with its "parent" lof strain.
7. Transfer a small plug from the rapidly growing "adapted" suppressor mutant from the 1 M KCl salt stress medium to normal CM (unstressed conditions). Cultivate the cultures for

3–4 weeks under stress-free conditions at 26 °C with light/dark 16/8 h, transferring the cultures every 8–10 days onto freshly prepared CM.

8. Then, transfer the suppressor strains back to the repeated salt stress (e.g., CM + 1 M KCl) and incubate them further at 26 °C with light/dark 16/8 h.
9. After a few days, the suppressor strains will be found to grow as fast as those taken directly from stress conditions. The phenotype appears to be stable and memorized (*see* **Note 5**).
10. Finally, before using the suppressor strains in other assays, take a single spore of the suppressor culture and propagate it to obtain a "pure strain" of one individually suppressed strain.

### 3.2 Further Verification of the Suppressor Strains

Please check two highly relevant parameters initially in the suppressor strains to avoid any possibility of contaminations or confusions about mixed cultures. Firstly, determine whether the genes originally inactivated in the "parent" lof mutants are still inactivated, for example, by means of standard Southern analysis. Additionally, point out whether the suppressor strains produce glycerol instead of arabitol as an osmotic stress response (for detailed procedure, *see* Chapter 4).

An example of the results of carbohydrate analytics for verification of suppressor strains of *M. oryzae* is shown in Fig. 2.

1. Impose hyperosmotic shock by adding osmotic stress stock solutions to the fungal strains, so that the final concentration in the culture broth is 0.5 M KCl.
2. Determine the intracellular levels of the major osmolytes (e.g., arabitol and glycerol) after 7 h (*see* **Note 6**).

## 4 Notes

1. It is important to know that it has not been possible for us to isolate suppressor strains from HOG pathway-independent osmosensitive *Magnaporthe* mutant strains, i.e., *ΔMostu1* (transcription factor in cAMP/PKA signaling pathway), *ΔMogpd1* (glycerol-3-phosphate dehydrogenase), or *ΔMoskn7* (response regulator protein), to date.

   Furthermore, long-term cultivation upon salt stress or other osmotic stress agents do not lead to a "better" osmoregulation or osmotolerance in *Magnaporthe* wildtype strains.
2. Fungal stocks are obtained from filter paper disks containing fungal mycelia. Stocks must be dehydrated in a desiccator for at least 2 weeks and stored in dry sterile tubes at −20 °C. The reason is that *M. oryzae* produces spontaneous mutations after a lot of subcultures. Therefore, the process of subculturing of

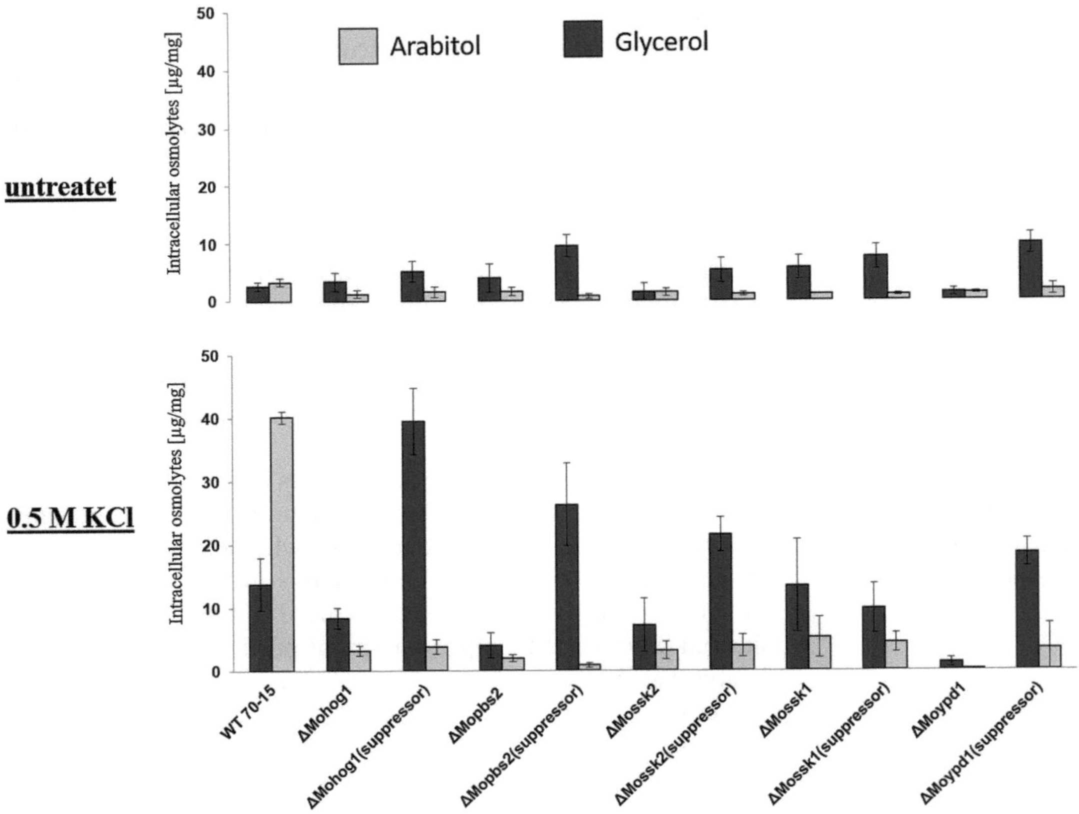


**Fig. 2** Carbohydrate production in the *Magnaporthe oryzae* wildtype strain (WT 70-15), the lof mutants, and the corresponding "suppressor" strains after 0.5 M KCl treatment. Compatible solute accumulation of glycerol (in dark gray) and arabitol (in light gray) was determined in mycelium after salt stress for 7 h. Carbohydrates were extracted and quantified by HPAEC-PAD analysis. Error bars represent the standard deviation of three biological replicates of each strain

this fungus should be avoided. Put the filter paper onto a CM agar plate and incubate at 26 °C with light/dark 16/8 h when using the fungal culture from stock.

3. The most effective concentration to obtain suppressor mutants should be determined previously in growth assays for each fungal species. Find the concentration in which fungal growth of the lof mutants with an inactivated HOG pathway is strongly restricted, whereas the wildtype strain should be affected but able to grow much better compared to the mutants.
4. The period of the suppressor event could vary between 4 and 12 weeks, whereby most of the events take place after 4–6 weeks. Do not transfer the cultures to freshly prepared stress medium several times; you have to leave them on the same plate for the whole time.
5. There will be some suppressor candidate strains which cannot cope with the osmotic stress. These strains should be discarded because the suppression is unstable.

6. Arabitol is the major intracellular compatible solute produced by the wildtype strain after osmotic shock. By contrast, the lof mutants *ΔMohog1*, *ΔMopbs2*, *ΔMossk2*, *ΔMossk1*, and *ΔMoypd1* could not produce either arabitol or glycerol in significant amounts. Interestingly, it was found that all suppressor strains *ΔMohog1(suppressor)*, *ΔMopbs2(suppressor)*, *ΔMossk2(suppressor)*, *ΔMossk1(suppressor)*, and *ΔMoypd1(suppressor)* responded to hyperosmotic stress by accumulating high amounts of glycerol rather than arabitol.

## References

1. Prelich G (1999) Suppression mechanisms: themes from variations. Trends Genet 15:261–266
2. Lewis D (1961) Genetical analysis of methionine suppressors in *Coprinus*. Genet Res 2:141–155
3. Morgan DH (1966) Suppression of 'purple' in *Coprinus lagopus*—an anomalous genetic situation. Genet Res 7:195–206
4. Hartman PE, Roth JR (1973) Mechanisms of suppression. Adv Genet 17:1–105
5. Hodgkin J, Kondo K, Waterston RH (1987) Suppression in the nematode *Caenorhabditis elegans*. Trends Genet 3:325–329
6. Adachi K, Hamer JE (1998) Divergent cAMP signaling pathways regulate growth and pathogenesis in the rice blast fungus *Magnaporthe grisea*. Plant Cell 10:1361–1374
7. Avery L, Wasserman S (1992) Ordering gene function: the interpretation of epistasis in regulatory hierarchies. Trends Genet 8:312–316
8. Jacob S, Foster AJ, Yemelin A, Thines E (2015) High osmolarity glycerol (HOG) signalling in *Magnaporthe oryzae*: identification of *MoYPD1* and its role in osmoregulation, fungicide action, and pathogenicity. Fungal Biol 119:580–594
9. Bohnert S, Antelo L, Grünewald C et al (2019) Rapid adaptation of signaling networks in the fungal pathogen *Magnaporthe oryzae*. BMC Genomics 20:763

# INDEX

## A



## B



## C



## D



## E



## F



## G



## H



## I



## J



## L



Stefan Jacob (ed.), *Magnaporthe oryzae: Methods and Protocols*, Methods in Molecular Biology, vol. 2356,
https://doi.org/10.1007/978-1-0716-1613-0, 

MIX
Papier aus verantwortungsvollen Quellen
Paper from responsible sources
FSC® C105338

If you have any concerns about our products,
you can contact us on
**ProductSafety@springernature.com**

In case Publisher is established outside the EU,
the EU authorized representative is:
**Springer Nature Customer Service Center GmbH**
**Europaplatz 3, 69115 Heidelberg, Germany**

Printed by Libri Plureos GmbH
in Hamburg, Germany